T0364173

Planet Management

NORTHWESTERN UNIVERSITY

MEDIA TOPOGRAPHIES

general editors   *James Schwoch*
Mimi White

Fernando Elichirigoity

PlanetManagement

Limits to Growth,

Computer Simulation,

and the Emergence of

Global Spaces

Northwestern University Press
Evanston, Illinois

Northwestern University Press
Evanston, Illinois 60208-4210
Copyright © 1999 by Northwestern University Press. Published 1999. All rights
reserved.

Printed in the United States of America

ISBN 0-8101-1587-5 (cloth)
ISBN 0-8101-1588-3 (paper)

Library of Congress Cataloging-in-Publication Data

Elichirigoity, Fernando.
    Planet management : limits to growth, computer simulation, and the emergence
of global spaces / Fernando Elichirigoity.
        p.   cm.
    Includes bibliographical references.
    ISBN 0-8101-1587-5 (cloth). — ISBN 0-8101-1588-3 (paper)
    1. Economic development—Social aspects.   2. Human ecology.   3. Manage-
ment.   I. Title.
HD88.E55   1999
304.2—dc21   98-43942
                                                                                  CIP

To Mary

Ahora más que nunca

Jose Arcadio Buendía spent several days as if he were bewitched, softly repeating to himself a string of fearful conjectures without giving credit to his own understanding. Finally, one Tuesday in December, at lunch time, all at once he released the whole weight of his torment. The children would remember for the rest of their lives the august solemnity with which their father, devastated by his prolonged vigil and by the wrath of his imagination, revealed his discovery to them: "the earth is round, like an orange."

—From *One Hundred Years of Solitude*
by Gabriel García Marquez

Contents

Acknowledgments

This book, which began as my doctoral dissertation, could not have been written without the support and encouragement of many people. Chief among them has been my advisor and friend, Richard Burkhardt, Jr. He has been unstinting in his support of this project, providing constructive criticism and encouragement throughout the whole process. Daniel Kevles read through an early version of the manuscript and provided supportive and much-needed advice on avoiding the pitfall of letting the theory obscure the substance of the research. Jim Schwoch, co-editor of the series within which this book appears, also provided excellent editorial comments and encouragement. Andrew Pickering encouraged me to think broadly and creatively and provided me with several useful insights. Geoff Bowker contributed generously of his time to read through the dissertation stage of the manuscript and made very helpful suggestions. Lillian Hoddeson provided me with excellent advice on doing research in archives and conducting interviews. Ted O'Leary contributed lucid explanations of theoretical points and offered many helpful insights on tackling interdisciplinary issues. Peter Neushul read a chapter of the manuscript and provided me with insights on postwar American history. To all of them goes my heartfelt gratitude. Any shortcomings in this work remain exclusively my own.

Anna Pignocchi, at the archives of Ente per le Nouve Tecnologie, l'Energia e l'Ambiente in Rome, gave me a great deal of assistance with the Aurelio Peccei Papers. Over many cappuccinos she also provided me with many insights on the functioning of the Club of Rome. Her help has been invaluable. At MIT, Helen Samuels, chief archivist, and Elizabeth Andrews, archivist, provided me with excellent professional support and friendly help in researching the archival materials. While at the California Institute of Technology, Ingeborg Sepp provided me with excellent administrative support. Sue Betz, at Northwestern University Press, expertly guided the manuscript through the last stages of revision and production.

Through the different stages of this work, I received generous financial and administrative support from many sources. The University of Illinois provided me with fellowship support and a travel grant. I was the recipient of the Tomash Fellowship from the Charles Babbage Foundation in 1993. These fellowships sustained me through the original research and writing period. The Fondazione Peccei from Italy provided me with a research grant and travel grant. They allowed me to study the Peccei papers in Rome. The California Institute of Technology honored me with a two-year postdoctoral fellowship in the humanities and social sciences division, from 1994 through 1996. It gave me ample and generous administrative support and time to transform the dissertation into a book.

This book could not have been written without the support of my wife, Mary Louise Eddy. Her sharp editorial comments improved the whole text immensely. Her love made it all worthwhile.

Chapter 3 introduces Jay Forrester, a pivotal figure in World War II and in the emergence of globality. Forrester, who began his career in the Servomechanisms Laboratory at MIT during the war, is the creator of System Dynamics, the set of modeling tools that was used to write the simulation model behind the *Limits to Growth* project. The chapter follows his career through his experience with the Servomechanisms Laboratory, the Whirlwind project, SAGE, and the Sloan School of Management at MIT. The chapter explores the way in which his experiences became embedded in the set of modeling rules that he would eventually call System Dynamics.

Chapter 4 covers the history of the Club of Rome, the organization of Western industrialists and nonelected high-level government bureaucrats that sponsored the *Limits to Growth* project. It also discusses some aspects of the life of Aurelio Peccei, one of the founders of the Club of Rome and the force behind that organization's impetus to focus on global problems. The overall aim of the chapter is to call attention to the emergence of new organizations that explicitly articulate a global viewpoint as opposed to a viewpoint based on the nation-state. The chapter also discusses the historically contingent ways in which the Club of Rome and Forrester came together.

Chapter 5 discusses early efforts by the Club of Rome to articulate novel conceptualizations about the interdependence of global problems which they called the "world problematique." It also discusses how Forrester was able to convince the club to adopt his modeling technique as the backbone of the World Problematique project, eventually to become the *Limits to Growth* project.

Finally, chapter 6 traces the history of the *Limits to Growth* project from formal approval to its public unveiling at a conference sponsored jointly by the Club of Rome, the Smithsonian Institute, and the Woodrow Wilson Center in Washington, DC.

One of the central concepts developed in this chapter is the historically contingent nature of the whole *Limits to Growth* project. Even though the project was to have a monumental impact, with over ten million copies of the report sold worldwide, its development was fragile and fraught with ambiguities and near-complete failure. In conclusion I argue that the *Limits to Growth* report, with its focus on thinking from the standpoint of the planet itself, should be understood as a seminal exemplar in the advent of the age of globality, the age of the global earth.

One

The Emergence of the Global Earth

> *In the middle of the 20th Century, we saw our planet from space for the first time. Historians may eventually find that this vision had greater impact on thought than did the Copernican Revolution of the 16th Century. From space, we see a small and fragile ball dominated not by human activity and edifice but by a pattern of clouds, oceans, greenery, and soils. Humanity's inability to fit its doings into that pattern is changing planetary systems fundamentally. This new reality, from which there is no escape, must be recognized and managed.*
>
> —*UN Commission on Environment and Development*

The past thirty years or so have witnessed a profound and pervasive change in the way we speak about, and the means by which we analyze, the interrelations between human activities and nature. Discourses and practices about these interrelations have shifted from conceptualizing nature as the passive backdrop for human activity and the bountiful fountain of resources for human activity to a view that considers human production as an intrinsic element of natural planetary processes.

The shift is signaled by the emergence of a number of discourses and scientific practices that center on the planet as a whole. The discourses consist of reports, studies, pronouncements, policies, and documents generated by private and public entities alike. I have called them "discourses of globality."

Concomitantly with these discourses, we have seen the development of a vast array of scientific practices devoted to the study of the planet as a whole. These "practices of globality" include, but are not limited to, satellite imaging and photographic surveillance of natural and anthropogenic phenomena at a planetary level, as well as computer modeling, simulation, and forecasting of the interactions of human production and the biosphere.

The total interactions and aggregations of the discourses and practices of globality make it possible, even "normal," to think of Earth as a total and irreducible system. In fact, we reconceptualize the whole planet as a field amenable to governmental, scientific, and managerial interventions. These interventions occur through myriad processes—negotiation of international accords, conducting scientific investigations, enacting governmental regulations, and undertaking military actions. This scientifically, politically, and culturally constructed field of intervention, based on the assumption of an ultimate and irreducible interrelation of human production and biosphere, is the "global earth."[1]

Discourses and practices of globality are dependent on, and mutually constitutive with, representations of global spaces and global perspectives. Probably the most generic, and starkly beautiful, of these representations is that of planet Earth photographed from outer space during the Apollo 8 space mission. The space image of Earth exercises a powerful influence on the imagination. In fact, statements are sometimes made to the effect that the image, by and of itself, is a transformative influence on human thought. The historian Donald Worster captures this sentiment when he asserts that

> the intuition of Columbus and his age that the world was a sphere has now become a photograph. . . . What that photograph says to us is elusive and contradictory. For some it seems to say that, at last, the earth is ours—we own it all, we dominate it. . . . For others, however, it says we live on a very small and vulnerable ball, the blue and green planet of life, floating alone and unique in the solar system. . . . Whatever the message in the photograph, we may not be the same after seeing it. Such a revelation of where we live may bring with it, slowly, at first almost imperceptibly, a revolution in thinking. (Worster 1987, 89)

For Worster, the possibilities of various interpretations of the image exist, but they are ultimately and unavoidably tied to the revolutionizing character of the image. Whether the new perception gained is one of total control or one about the fragility of life, and about the consequent need to conserve and preserve, the view of Earth from outer space remains a source for new understandings.

The connection of the picture of Earth from outer space with hope for its revolutionary potential for peace and understanding has been shared by people from many walks of life. President Lyndon B. Johnson, in one of his last acts as president, sent a copy of the color picture of Earth, taken by astronauts of the Apollo 8 mission, to every head of state, including Ho Chi Minh. The context of Johnsons's act suggests that he wanted to show that every member of the human species was in the same boat (Unger and Unger 1988).

The image of Earth from outer space has become a ubiquitous symbol. It is, for example, often and extensively used by environmental groups from all over the world. The appropriation of the image for environmental ends is perhaps a bit ironic if one thinks that the techno-scientific structure behind the taking of that picture was not built with particular care for the environment. As the historian Charles Bergman notes, at least one species of bird in Florida, the dusky seaside sparrow, was hurried into extinction through destruction of its marshland habitat by the space program.[2]

The notion of the new perspective as a source of new possibilities was also suggested by Frank White, who was in 1987 a senior associate at the Space Studies Institute. He called the new perspective the "overview effect." He stated that this perspective was first experienced rather naturally by astronauts as they looked to Earth from outer space. White quotes astronaut Russell Schweickart of the Apollo 9 mission to describe this effect:

> When you go around the earth in an hour and a half, you begin to recognize that your identity is with that whole thing. That makes a change. You look down there and you can't imagine how many borders and boundaries you cross, again and again and again, and you don't even see them. There you are—hundreds of people in the Mideast killing each other over some imaginary line that you are not even aware of that you can't see. (White 1987, 12)

The overview effect, according to White, is something to be promoted because of the new possibilities it opens for seeing the world. He suggested that

> hearing an astronaut speak, seeing a film, or looking at a poster of the whole Earth begins the adoption process by bringing awareness of the overview to the audience. These experiences are not as deep as being in space, but the impact is broader because a film or poster can be replicated more easily and less expensively than the experience itself. (White 1987, 70)

Elsewhere in his book White suggests that the overview effect will gain greater solidity as the image of the earth from outer space is communicated more frequently by the mass media.

Written in 1987, White's work had a clear ideological purpose: to lend support to the space program. As a member of the Space Studies Institute, White was well aware that the space race was less animated by narratives of national competition than it had been in an earlier cold war epoch and thus was less supported by public moneys. White, like other space exploration boosters, was trying to find a new narrative to make clear the need for further investment in the space industry. The narrative of a joint universal effort in the name of a common humanity and a common home, that is, of the planet Earth itself, is White's way of rationalizing support for the space program. He quotes approvingly from *Omni* magazine: "The meek shall inherit the earth. The rest of us will go to the stars" (White 1987, 117).

The more immediate ideological intent of White's book should not distract us from seeing the claims about the overview effect as a specific example of a discourse of globality. It embodies some of the characteristics of such discourses, including a global view of the planet and a need to conceptualize human activity from that perspective. It also illustrates the ocular language that is embedded in discourses of globality, namely, the viewing of Earth from outer space detached from reference to human history.

Global ocularity gains a more concrete reality when it becomes connected to yet other discourses and practices. The way this comes about can be explored by looking at a special issue of *Scientific American*, a magazine of general circulation but aimed at a scientifically literate reader.

As its title indicates, the September 1989 issue of *Scientific American* is entirely dedicated to "Managing Planet Earth." The cover of the magazine is as interesting as the focus of the text inside. It presents a painting depicting a perspective of Earth from outer space. The painting represents the impact of human activity against the dark contours of the different continents, depicting it as a multitude of light dots. These dots, in turn, represent, in the aggregate, such different types of human activity as energy generated by cities, slash-and-burn agriculture, natural gas flares, and grassland burning. The data used to provide a scientific basis for this artistic interpretation were culled from satellite photographs made by the Defense Meteorological Satellite Program of the U.S. Air Force.

The contents of this special issue offer a collection of articles on different aspects of the impact of human activity on the planet as a whole. Articles and photographs present humanity in the aggregate, though distinctions between the rich and the poor are made when talking about population growth. One of the articles, written by Stephen Schneider (1989), then head of the Interdisciplinary Climate-Systems Program at the National Center for Atmospheric Studies, deals with the threat of global warming, now intensifying, according to Schneider, due to ever-rising concentrations of carbon dioxide and other gases. Another article, written by E. O. Wilson, a Harvard University entomologist best known as the main founder of sociobiology—a fact not mentioned in the magazine—worries about the rapid loss of biodiversity. According to Wilson, "habitat destruction, mostly in the tropics, is driving thousands of species each year to extinction. The consequences will be dire unless the trend is reversed" (Wilson 1989, 108). Wilson is especially worried about the loss of many genetic codes that could be useful to biologists in the future.

Yet another article deals with the problem of population growth. This article, written by Nathan Keyfitz, professor emeritus at Harvard and head of the population program at the International Institute for Applied Systems Analysis (IIASA), argues that population growth in poor nations hastens the degradation of the environment and threatens the very economic development that these countries want and need (Keyfitz 1989). The magazine also has a set of articles that could be thought of as "solutions" to—or at least possible ways of alleviating—the problems described in the first set of articles. The titles of this second set of articles suggest a generalized technocratic orientation: "Strategies for Energy Use," "Strategies for Sustainable Economic Development," and "Strategies for Manufacturing." The last article, by Robert Frosch, a theoretical physicist at the General Motors Research Laboratories, and Nicholas Gallopoulos, a chemical engineer also at General Motors, argues for the need to think of industrial production moving toward a model they call the "industrial eco-system"—a system to be characterized by decreasing use of materials and closed-system manufacturing.

The articles that comprise the magazine are lavishly illustrated. These illustrations include the usual *Scientific American* diagrams and many color pictures of the environment in different stages of stress. Also included are several satellite pictures, sometimes digitally enhanced, and computer-generated images depicting the results of computer simulation runs. The message throughout is that all manner of human concerns is dependent upon the state of the whole planet.

This issue of *Scientific American* also carries a heavy dose of advertising. One advertisement—for a cleaning powder—talks about this product's contribution to keeping planet Earth clean for over 100 years. It includes a picture of Earth from outer space. Other ads, many for global corporations, link their products with helping the environment, or connect them more generally with nature in a positive way. An ad for Boeing Corporation, for example, links, via drawings and text, the design of a Boeing jet plane with the aerodynamic design of birds and insects. Amoco Corporation has an ad depicting their plastic recycling program.

Not every ad is keyed to the overall theme. While many of the ads echo the theme of the special issue, several of them make no conscious connection to the environment and simply celebrate consumption of their products, which, like BMW cars, can be had with the help of a healthy purse. In general, though, there is a conscious effort to link the advertising to planetary concerns in some way.

This issue of *Scientific American* is, in short, an exemplar of the heterogeneous array of discourses and practices of globality which, in their aggregate, construct a field of action and intervention. The cover alone brings together many of the elements that suggest the global space, while the title, "Managing Planet Earth," conveys ways of looking at Earth and what happens within it in a peculiar way—in a managerial way. Earth ceases to be the backdrop to human history and culture and becomes the all-encompassing system that needs to be managed to attain optimum performance.

The cover of the magazine also embodies several aspects of a new regime of ocularity or vision. It assumes a real or simulated "mental eye" cast from some point in outer space toward Earth. This ocularity is generally produced by machine vision (the exception being the astronauts' photography) and is usually analyzed and enhanced by digital processes. In fact, the ensemble of practices built around machine vision, such as satellite imaging and scanning as well as digitized interpretation, are constitutive of the new regime of ocularity that accompanies the global earth.

The articles inside the magazine are largely dependent on, and make constant reference to, this new ocularity of machine vision. The new ocularity is connected, via the text, with some form of systems analysis. This mode of analysis, which I explore in more detail below, facilitates the discussion of Earth as a system that is composed of both organic and inorganic, as well as natural and technological, subsystems, amenable to the same rules of analysis.

Another aspect of the mental and material technologies that constitute the discourses and practices of globality represented in this issue of *Scientific American* is the use of computer modeling and simulation. Schneider, for example, in his article on global warming, makes reference to a climate model that he and another scientist constructed to project surface temperatures of Earth with an atmosphere containing twice the present level of carbon dioxide. The calculations of the computer model, in turn, were transformed into graphic simulations that show some of the effects of global warming on the surface temperature of Earth (Schneider 1989, 76).

Modeling and simulation are important components of the emerging practices of globality. They present a fundamental epistemological shift from representation of the natural world to a simulation of the natural world. The simulation runs of Schneider's modeling serve as a source of scientific facts about the health of the planet. Since the basis of the argument is the potential for future catastrophe, the traditional experimental procedures are less able to stabilize scientific truth because traditional experiments cannot measure what has not yet happened. In fact, much of the current discourse of global warming is largely built around computer modeling. Modeling brings together disparate data from many sources—such as satellite imaging, local studies of change in vegetation, and human consumption of energy at the micro and macro levels—and provides a coherent narrative for the presentation of the data.

The issue of *Scientific American*, as a discourse, is a text with a particular set of ideas running through it. It is a text, however, that makes constant reference to, and is dependent on, a set of heterogeneous practices, such as computer simulation and modeling and satellite-based machinic vision.

A paramount practice, pervasively connected with the discourses of globality, is satellite surveillance. This practice is embedded in the text, cover, and illustrations of the magazine under discussion. The mutually constitutive nature of discourses and practices can be seen in that the new practices themselves generate discourses justifying and advancing the new practices. The new regime of ocularity begins to take shape in precisely this thickening interconnection of discourses and practices, which, in turn, attracts and connects with other, formally separate, domains of knowledge.

It is important to stress the radical nature of this shift to modeling, simulation, and manipulation of machine vision products like satellite imaging. They constitute a new mode of representation and visuality. As Jonathan Crary argues:

> The formalization and diffusion of computer-generated imagery heralds the ubiquitous implantation of fabricated visual "spaces" radically different from the mimetic capacities of film, photography, and television. These latter three, at least until the mid-1970s, were generally forms of analog media that still corresponded to the optical wavelengths of the spectrum and to a point of view, mobile or static, located in real space. Computer-aided design, synthetic holography, flight simulators, computer animation, robotic image recognition, ray

tracing, texture mapping, motion control, virtual environment helmets, magnetic resonance imaging, and multispectral sensors are only a few of the techniques that are locating vision to a plane severed from a human observer. (Crary 1990, 1)

The significance of this radical shift in the standpoint of vision is that it represents a move away from what the embodied human eye sees to what disembodied machine systems "see." It is not simply that the human eye is enhanced by machines, but that the ways machines "see"—through ultraviolet and infrared technologies, through simulation of elements unseen by the human eye—becomes increasingly important. Thus, vision becomes increasingly disembodied, no longer dependent on the human body for its existence. This disembodiment is, of course, not unique to vision. The use of computers, for example, allows for the displacement of some forms of intelligence, as is the case with the growing use of "expert systems" in several professional fields such as nursing, bank lending, and legal practice. These systems aim to capture in formal rules the explicit and tacit knowledge of professionals and to embody them in software that can be used by people without particular expertise in a given professional field.

Another illustration of the increasing interconnections of discourses and practices of globality with a new regime of ocularity can be found in the Mission to Earth program sponsored by NASA (Baker 1990). The goal of this program, as James Baker reports it, is

> to achieve a scientific understanding of the entire Earth system on a global scale by describing how its component parts and their interactions have evolved, how they function, and how they may be expected to continue to evolve on all time scales. The challenge: to develop the capability to predict those changes that will occur during the next decade and the next century, both naturally and in response to human activity. (Baker 1990, 115)

The plan developed by NASA was based in part on an earlier set of recommendations of the Committee on Science, Engineering, and Public Policy of the National Research Council. This committee advanced the need for more satellite surveillance of our planet. Their recommendations read in part:

> To advance our understanding of the causes and effects of global change, we need new observations of the Earth. These measurements must be global and synoptic, they must be long-term, and different processes

such as atmospheric winds, ocean currents, and biological productivity
must be measured simultaneously. We have learned that major advances
in the Earth Sciences have come from syntheses of new ideas drawn from
such global synoptic observations. (Baker 1990, 116)

The structure of Mission to Earth will encompass three main elements, some now in the process of implementation. Baker describes these
elements as

global long-term measurements from satellites and ground-based systems
to document the physical, chemical, and biological processes responsible
for the evolution of Earth on all time scales; the use of that data in
quantitative models of the earth system to identify and simulate global
trends and to make predictions; and the establishment of an information
base for effective decision-making in responding to the consequences of
global change. (Baker 1990, 115)

As we can see from the above quotations, the Mission to Earth program is
an all-encompassing apparatus for the synoptic surveillance of the planet.
The processes of satellite imaging at one end and computer simulation
at the other end constitute the frame of a cybernetic panopticon of
immense proportions. The system has been enhanced in phase 2 of the
project with the launching of a dedicated satellite, in 1998, that generates
integrated measurements of the earth's processes. These measurements
will become the backbone of a fifteen-year environmental database of
immense proportions.[3]

While this NASA project is one of the most comprehensive, it is not
by any means the only one. The United Nations has in the works a global
database with functions similar to those of the NASA program (Gwynne
and Moneyhan 1989).[4] The Global Resource Information Database, as it
is called, is envisioned as one that "will provide data and information to
scientists, planners, and decision makers in their job of managing and
assessing human impacts on critical environmental issues" (Gwynne and
Moneyhan 1989, 250).

One of the interesting rationales provided in support of this project
is the need for new forms of information (based on digitization), distinct
from those of the past. As Michael Gwynne and Wayne Moneyhan state:

Traditional access to environmental data—in shelves of reports and
proceedings as well as in fast-aging maps and charts—no longer meet
the demands of planners faced with a world in which the nature of
environmental change is infinitely complex. With the development of

computers that can handle large quantities of data, a global database is now possible. (Gwynne and Moneyhan 1989, 250).

Particularly significant in the above statement, and exemplary of discourses connected to planet management, is the demand for global databases. Such databases, used to collect data about the planet, are also constitutive practices of the global earth. The new ways of seeing and calculating that have been spawned by focusing on the planet as a whole call for new ways of inscription and manipulation of information. "Aging maps and charts" easily interpreted by the human eye now give way to the digitization of the new information, making them amenable to machinic manipulation. Databases provide the locus for stored new information. Such information is not immediately available to the human eye. It must first be mediated by machinic techniques, such as modeling and simulation. This use of information, in turn, produces knowledge and objects of knowledge that could not exist without that particular technology. The increasing use of databases, whether centralized and located in a specific machine or distributed over a large number of computers and connected via the Internet, is radically changing our notions of what information is and of what can be done with it.

Yet another example of discourses of globality is a 1991 report published by the Trilateral Commission entitled *Beyond Interdependence: The Meshing of the World's Economy and the Earth's Ecology* (MacNeill, Winsemius, and Yakushiji 1991). The Trilateral Commission is a group of top politicians, corporate heads, and bureaucrats drawn from what political scientist Stephen Gill has called the "international establishment" (Gill 1990, 155). *Beyond Interdependence* is not particularly unique; there are many other examples of similar works. In fact, the main author, Jim Mac-Neill, was secretary general to the World Commission on Environment and Development when *Our Common Future,* from which the quotation at the beginning of this chapter is taken, was commissioned and published. The main significance of this report for my purposes lies in its being an exemplar of the discourses that, in their aggregate, constitute the new spaces of governmentality that I have called the global earth.

The report starts by making reference to the speed of change and to technoscientific practices connected to that change. It states:

> Events are accelerating on several fronts simultaneously—economic, ecological, and political—and are forcing profound changes in the relationships among peoples, nations, and governments. . . . Many and new emerging technologies in biology, materials, construction, satellite monitoring, and other fields offer great promise for increasing the

production of food, developing more benign forms of energy, raising
industrial productivity, conserving the earth's basic stocks of natural
capital, and managing the environment. With global communications
and ever greater access to information, people can now begin to exercise
responsibility for every part of the planet. (MacNeill, Winsemius, and
Yakushiji 1991, 3)

The above quotation is, again, characteristic of the new discourses of
globality. It makes reference to growing speed, magnitude, and complexity. This reference tacitly embodies what is more explicit in other parts of
the report: a call for better techniques and strategies of management that
can deal with vast changes. The quotation also refers to new technologies, including satellite monitoring and global communications. These
technologies figure prominently in the constitution of global spaces. As
considered in more detail below, discourses that allude to the global earth
often make reference to these technologies.

The report makes repeated reference to the global environment
and its irreducible connection to human production. On page 4, we read
that

the world has now moved beyond economic interdependence to
ecological interdependence—and even beyond that to an intermeshing
of the two. The earth's signals are unmistakable. Global warming is
a form of feedback from the earth's ecological system to the world's
economic system. So is the ozone, acid rain in Europe, soil degradation
in Africa and Australia, deforestation and species loss in the Amazon. To
ignore one system today is to jeopardize the others. The world's economy
and earth's ecology are now interlocked. (MacNeill, Winsemius, and
Yakushiji 1991)

These new global spaces, organized around the biosphere-humanity system on the one side and the global economy on the other, have emerged
largely out of novel reconceptualizations or "redistributions" of nature
and of information. In the new conception of nature, perhaps best
illustrated by the notion of "sustainability," the relations between human production and the planet have been radically altered. In the past,
nature, when thought of at all, was the unproblematic source of raw
materials for production and the uncomplaining ground for dumping
the toxic by-products of those production processes. This view of nature
has undergone rapid and radical transformation. Human production
and consumption are increasingly perceived as integral parts of nature,
as planetary processes.

From the standpoint of the older discourses and practices, nature has been redistributed. It no longer occupies specific and well-delineated spaces separate or distinct from human activity. Nature is now intimately involved in that activity. From the standpoint of the emerging discourses and practices, human activity is reintegrated into the larger planetary systems with the effect of deeply transforming the aims and methods of that activity.

Planet management must be understood in terms of an altered view of nature. It must also be understood in terms of the emergence of the concept of "biosphere"—a concept central in linking global spaces with systemic notions of the interrelationship of all life. Because the concept encompasses the physical dimensions of Earth, "biosphere" makes inescapable our dealing with the global nature of certain problems. The following section traces the use of the term "biosphere" and the historical transformation of its meaning.

Examining the shifting views of biosphere provides us with another "take" or "snapshot" of the processes involved in the emergence of the global earth as a field of intervention. The evolution of the use of the term "biosphere" may be seen as we contrast its use by Vladimir Vernadsky, a Russian scientist who was the first to provide a detailed theory of the biosphere, and its different use by James Lovelock, who talks of biosphere in terms of his controversial notion of "Gaia." The contrast illustrates the radical reconceptualizations of the relationship between humans and nature, so central to the discourses and practices of globality, that have emerged in the last forty years.

For most of his scientific life, Vernadsky, a Russian scientist born in 1863, was concerned with geology, crystallography, and chemistry. In the last twenty years of his career, however, he became increasingly interested in the relations between biological and geological processes which he first explored in 1917, by focusing on the concept of the biosphere (Vernadsky 1929).

This concept was first proposed by an Austrian geologist, Eduard Suess, who in an 1875 book on the Alps used the word "biosphere" to describe the life-supporting envelope around primordial Earth. Vernadsky was much taken with Suess's concept. As he stated in 1924: "In advancing the new notion of a particular terrestrial envelope, determined by life, Suess enunciated, in reality, a new empirical generalization of great import, for which he had not foreseen all the consequences."[5] Vernadsky proceeded to explore the consequences of the idea of the biosphere in many writings, most particularly with his book *Biofera* (first published in Russian in 1924, and later translated into French in 1929), and in many articles in scientific journals.

The biosphere, as posited by Vernadsky, is that region of the planet where life, or more precisely, living matter exists and where solar radiation interacts with Earth and transforms it. Living matter, distinct from inert matter, is the sum of living organisms. While maintaining clear boundaries between inert and living matter, Vernadsky posited a constant interrelationship between these two forms of matter. The mechanisms of interrelationship are to be found in the continuous biogenic migration consisting of such life functions as breathing, nutrition, and reproduction—all of which facilitate the migration of inert matter to living matter—and in organic decay after living organisms die, which facilitates the migration of atoms from living to inert matter.

It is important to highlight that Vernadsky was strongly committed to the integrity of each and every individual organism composing living matter. He posited the biosphere as a complex relation of organisms and inert matter, but this interrelation was not dependent on erasing the boundaries between the different living organisms and between living and inert matter. At the same time he hypothesized that no living organism existed in a state of radical freedom, unconnected to the biosphere. Thus, man, in Vernadsky's eyes, is irrevocably tied to the biosphere through a complex set of interrelations affecting all living organisms. This sense of interrelations is, as Alexej Ghilarov reminds us, a continuation of the notion of the "underlying causal unity of Nature" that is at the core of the modern episteme that first emerges at the end of the eighteenth century (Ghilarov 1995, 199).

Dependence of man and the biosphere is to be understood, according to Vernadsky, within a Darwinian evolutionary process that made man the highest representative of that process. He saw the evolutionary process as culminating in the human brain, which provides man with the intelligence necessary to transform the biosphere. The process of evolution culminates in man's ability to control that biosphere for his purposes. The biosphere exists for the benefit of man. The fact that man is not an organism independent from his environment is seen positively by Vernadsky as scientific grounding for the achievement of man's dominion over the biosphere as a whole. The possibility of man's being in charge of the biosphere highlights Vernadsky's commitment to the ultimate integrity of the human species and, by extension, to the discourses and practices associated with the sign of "man." These discourses are normally connected with the Enlightenment.

A more recent conceptualization of "biosphere" may be meaningfully contrasted with that of Vernadsky—the one proposed by James Lovelock in his Gaia hypothesis. Lovelock conceived of the Gaia hypothesis in the early 1960s while he was working at the Jet Propulsion Laboratories of the California Institute of Technology (Lovelock 1979). His specific task there was to propose experiments that could help the unmanned Mariner expedition to Mars determine whether the planet had life forms in it. Lovelock collaborated with Dian Hitchcock, a philosopher who was involved in determining the logical content of the different experiments proposed. They concluded that the most certain way of determining life on a planet was to study its atmosphere. They reasoned that the metabolic functions of organisms would change the atmosphere and disturb any state of chemical equilibrium. They discovered that while Earth's atmosphere was in a "persistent state of disequilibrium," that of Mars was very near total equilibrium (Lovelock 1988, 6). The discovery did not endear them to NASA bureaucrats, who needed justifications to convince Congress to finance the project. The discovery, however, led Lovelock to continue exploring the fact that, even though Earth's atmosphere is in a state of disequilibrium, this state never collapses into chaos. Indeed, it had been maintained for literally billions of years in a stable state. Lovelock describes this process:

> The earth has remained a comfortable place for living organisms for the whole 3.5 billion years since life began, despite a 25 per cent increase in the output of heat from the sun. The atmosphere is an unstable mixture of reactive gases, yet its composition remains constant and breathable for long periods and for whoever happens to be the inhabitant. This, and other evidence that we live in the "best of all possible worlds," was the basis of the Gaia hypothesis; living organisms have always, and actively, kept their planet fit for life. In contrast, conventional wisdom saw life as adapting to the otherwise inescapable physical and chemical changes of its environment. (Lovelock 1986, 25)

The notion that the planet has been maintained in a homeostatic state for billions of years by life itself led Lovelock to claim that Earth is a living organism. Lovelock argues against dismissal of the idea with the following analogy:

> Could a planet, almost all of it rock and that mostly incandescent or molten, be alive? Before you dismiss this notion as absurd, think, as did the physicist Jerome Rothstein, about another large living object: a giant redwood tree. That is alive, yet 99 per cent of it is dead wood. Like the Earth it has only a skin of living tissue spread thinly at the surface. (Lovelock 1986, 25)

Lovelock's notion has been strongly criticized. Indeed, for the first fifteen years after announcing it, most scientists dismissed it out of hand.

It was only in 1988 when the climatologist Stephen Schneider, himself a partial critic of the Gaia hypothesis, organized a conference on the subject that the Gaia hypothesis was discussed more seriously. While there was initial resistance to the hypothesis, in the end "conferees gave Lovelock a standing ovation. Even critics said he had developed an ingenious way of looking at the world" (Lyman 1989, 56).[6]

One of the reasons that Lovelock's theory had been dismissed, and still is, by many scientists is that they see in it a teleological element (Schneider 1992). They perceive Gaia as somehow a conscious being that "thinks" to maintain the creative disequilibrium of the atmosphere within conditions amenable to life. Lovelock strongly argues against any teleological understanding of his theory. He developed a computer model of an imaginary planet named Daisyworld to illustrate the possibility of homeostasis without needing any teleological mechanism. With this model Lovelock shows that a planet with white and black daisies regulates its temperature within a relatively normal range when it is postulated that colder or warmer weather encourages the growth of more black daisies (which, in turn, absorb heat) or white daisies (which reflect some heat back to outer space). According to the model, things work even better when more varieties of daisies are introduced into the computation. He concludes that planetary homeostasis is maintained in this model without recourse to conscious governing principles (Lovelock 1988).

The notion of biosphere embedded in the Gaia hypothesis differs from Vernadsky's notion of biosphere in many ways. An essential difference is the relation of man to Gaia. Lovelock, unlike Vernadsky, does not see that the evolution of life favors man in any way. Gaia, as Lovelock sees it, is a cybernetic system with homeostatic tendencies that does not favor any one of the organisms that compose life. The system is geared toward maintaining conditions for life itself, rather than maintaining the life of any specific organism, such as man. As Lovelock puts it, "in Gaia we are just another species, neither the owners nor the stewards of this planet" (Lovelock 1988, 14).

Another aspect of the Gaia hypothesis that sharply differentiates it from Vernadsky's notions of the biosphere is the way in which homeostasis is maintained. In Vernadsky's conception the tacit idea is that the biosphere is evolving toward a more perfect state of balance and man's control over it will only make things better. Lovelock, on the other side, posits Gaia as a complex feedback system able to maintain large oscillations in the conditions that make balance possible. However, if the oscillations become too large, Gaia will abruptly readjust to a new stable state. He states: "Gaia theory predicts that the climate and chemical composition of the Earth are kept in homeostasis for long periods of time until some internal contradiction or external force causes a jump to a new stable state" (Lovelock 1988, 13). According to Lovelock, there are absolutely no guarantees that the new state will maintain conditions of life amenable to humans. It can only be asserted that Gaia will maintain conditions for life itself and the organisms that adapt to the new circumstances. This notion no longer locates man at the center of the biosphere. Man is just another species trying to get by.

It is important to highlight at least three central elements in Lovelock's hypothesis. They are: (1) a view of planet Earth from outer space, tacitly dependent on satellite imaging and surveillance; (2) the use of computer modeling to embody that view and test different hypotheses on the possible outcomes of the behavior of the Gaia system when certain variables, such as increased pollution, are introduced; and (3) dependence on complex feedback systems theory. These material and mental technologies made it possible for Lovelock to think about the biosphere in new ways. For example, the possibility of modeling Earth as a total system with human activity as a single aggregated variable among many loosened his dependence on older discourses, like those within which Vernadsky operated—discourses that gave man a more central position. Similarly, the possibility of thinking of Earth as an irreducible system composed of many subsystems facilitated his breaking away from scientific discourses that were dependent on maintaining sharp boundaries between the organic and the inorganic or between the human and the technological.

The modeling that Lovelock performed embodied in some ways the view of Earth from outer space. Lovelock himself suggests that this standpoint of ocularity opened up new possibilities. He writes:

> The real bonus has been that for the first time in human history we have had a chance to look at the Earth from space, and the information gained from seeing from the outside our azure-green planet in all its global beauty has given rise to a whole new set of questions and answers. (Lovelock 1979, 8)

While Lovelock's conception of the biosphere may represent an extreme within the continuum of the discourses and scientific practices of globality, it is not unique in displacing man from the center of the concept. Martin Price, a researcher at the National Center for Research, notes after looking at conceptualizations of the relation between human production and the biosphere that there has been a steady shift from an anthropocentric position in the 1970s to what he calls a geocentric one in global change studies (Price 1989). He locates the first shift in a series of studies on the relations between biogeochemical processes and human activities sponsored by NASA in 1982. Price calls those studies "holistic" because they included natural processes as part of the equation. He traces the shift to "geocentric" thinking to a study sponsored by the International Council of Scientific Unions in 1983 entitled "Toward an International Geosphere-Biosphere Program: A Study of Global Change." Price characterizes the progression from anthropocentric to geocentric by stressing that the anthropocentric model looks principally at the relations between people and their institutions. On the other hand, the geocentric approach centers on natural processes, first included in more holistic studies, and sees humanity as just one of many factors in the functioning of the system that we call planet Earth.

The move from an anthropocentric framework of understanding to one that centers on the biosphere as a whole has a history with many distinct and sometimes unconnected paths of development. Most of those paths started or gathered strength out of the crucible of World War II. We now turn to that story.

Notes

1. While I will not discuss it in this book, it should be mentioned that the massive array of discourses and practices constituting "global earth" is, in turn, augmented by another set of discourses and practices that center on the globalization of production and consumption, what is commonly known as "the global economy." The focus of this work is on the concepts underlying the notion of the global as a space of activity, rather than on the content of those activities.

2. Bergman further states that "the swampy marshes around Merritt Island were flooded to destroy the mosquitoes plaguing the development of the Kennedy Space Center and nearby Titusville. The high waters, created by impoundments, destroyed the marsh-grass habitat of the sparrow. Then, in order to make commuting more convenient for the space center workers, the Beeline Expressway was built between Orlando and Cape Canaveral, cutting right through the middle of one of the birds' marshes" (Bergman 1990, 51).

3. For more details, see NASA's web page on the project at http://www.hq.nasa.gov/office/mtpe.

4. For a discussion of several examples of global databases focusing on the Earth as a whole, as well as the myriad technical and managerial issues connected with them, see Mounsey and Tomlinson (1988).

5. My translation; the original quote from the French reads: "En établissant la nouvelle notion d'une enveloppe terrestre particulière, déterminée par la vie, Suess énonçait en réalité une nouvelle généralisation empirique d'une grande portée, dont il n'avait pas prévu toutes les conséquences" (Vernadsky 1929, 93).

6. A published version of the proceedings of that conference as well as other articles pro and con the Gaia hypothesis can be found in Schneider (1992).

World War II and the New Technologies of Knowledge

We leave history to enter simulation.

—*Jean Baudrillard* [1]

World War II had profound effects on the production of knowledge. Practices that either first originated during the war or, having originated before, were more systematically developed came to influence greatly the landscape of postwar America, and by extension, of much of the Western world.

Out of the war effort emerged many new technologies, such as radar, computers, and lasers. We could call these "material" technologies. The war effort also brought forth what could be called "mental" technologies, such as OR, cybernetics, and systems thinking. Several of these mental technologies, in turn, became transformed, after the war, into novel management techniques, either through extension, as in the case of OR, or through modeling, as in the case of System Dynamics.[2] What follows is a discussion of these new mental technologies and their contributions to the emergence of globality.

The New Mental Technologies: The Case of OR

> Undaunted, the Commander picked the appropriate "search plan" and executed its complicated gyrations throughout the night. After eleven hours of geometrical wanderings, the submarine was picked up almost dead ahead! After prolonged depth charging, a deep rumbling explosion

underwater and the strong smell of Diesel oil testified to at least the temporary end of another enemy mission. (Steinhardt 1946, 650)

The description quoted above, of an antisubmarine operation by an Allied ship during World War II, was written by Jacinto Steinhardt. Steinhardt was a chemist by training and one of the original members of an OR group set up in May 1942 by the National Defense Research Committee (NDRC). This particular group, under the direction of physicist Philip Morse, was set up to help the U.S. Navy deal with the constant threat that Nazi U-boats posed to Allied shipping in the Atlantic Ocean corridors.

The antisubmarine measures that this and other similar OR groups explored were divided into three broad categories: (*a*) aircraft search, (*b*) ship convoy protection, and (*c*) antisubmarine depth bombs workings. These groups were engaged in interdisciplinary quantitative analysis to optimize the ability of Allied forces to move ship convoys to Europe, mainly from the United States, and to eliminate, or at least weaken, enemy operations against those convoys. The group in charge of optimization of aircraft search, for example, engaged in mathematical analysis of such variables as total ocean area to be scanned, speed of planes, and probability of sightings, as well as ergonomic studies on the effect of monotony of task on personnel. These and other variables were sifted through complex probability studies until an optimum flight pattern was established for a specific geographic area (Morse and Kimball 1951). The application of OR to enemy submarine search transformed the process from what Steinhardt calls "general search" (just going out for a look around) to "planned search." The end result of these procedures was a series of "scientifically designed flight plans" (Steinhardt 1946, 651–52).

Similar work was done in studying the optimum size of a ship convoy, as well as the optimum length of the fuse of depth bombs used against U-boats. In the case of the ship convoys, it was decided, counterintuitively, to actually increase them in size from forty to 100 ships. This decision dramatically improved the safety of convoys. The study on the optimum length of depth bomb fuses led to an improved "hit" rate of about 20 percent.

The instances of OR undertaken during the war by the group under Morse were multiplied many times by other groups in both Great Britain and the United States. What follows is a brief discussion of OR on both sides of the Atlantic. Operations Research emerged from the very specific and concrete needs of the war, requiring that scientists and military personnel learn to interact with one other. This learning, repeated throughout the war across the full spectrum of technoscientific practices, has left a profound mark on the postwar era and was certainly present, as I discuss below and in the next chapter, in the development of such "mental" technologies as Forrester's System Dynamics.

The origins of OR are not overly clear. Part of the opaqueness is due to the very vagueness of what actually constitutes OR. Furthermore, some of the changes that later came to be identified with specific OR practices originated as small changes within mundane tasks, and the record of those changes has been lost in the lack of recognition of their possible importance.

Florence Trefethen, who wrote one of the earlier historical treatments of OR, states that OR started to be practiced in a systematic manner in Great Britain at the outbreak of World War II. In 1939 some ad hoc groups within the Royal Air Force began to use the knowledge of civilian scientists for the purpose of radio location (Trefethen 1954). These studies were complemented with an analysis of the ways of integrating the "newly developing radar system of early warning against enemy air attack with the older system of operational control based principally on the Observer Corps, whose members were trained in the sighting, identification, and reporting of planes" (Trefethen 1954, 5). Such new techniques developed as the air force moved away from reliance on the human eye and toward machinic vision, in this case, radar, for seeing.

In August 1940, the antiaircraft command of the British army organized a group of scientists to study ways to better integrate the newly developed radar system with existing antiaircraft weaponry. The group was established under the direction of P. M. S. Blackett, an astrophysicist who had earned a Nobel Prize for his studies of cosmic rays and who was later dubbed the "father of Operations Research" by James P. Baxter, who was the official historian of the Office of Scientific Research and Development (OSRD) during the war (Baxter 1946, 404). The group, which became known as Blackett's Circus, was initially composed of a variety of scientists from several disciplines. They included three physiologists, two mathematical physicists, one astrophysicist, one army officer, one surveyor, one general physicist, and two mathematicians (Trefethen 1954). This pattern of multidisciplinarity was to be repeated in all OR groups that were set up by the Allies during the war. The group headed by Morse within the U.S. Navy, for example, was originally composed of six mathematicians, fourteen actuaries, eighteen physicists, three chemists, two biologists, and one architect (Baxter 1946).

These interdisciplinary groups characterize OR work and can be found in many other areas of technoscientific work during the war.[3] The success of these interdisciplinary collaborations goes a long way to explain the rise of such theoretical apparatuses as cybernetics and systems thinking, which did away with the distinctions between the organic/natural and the technical and which gave confidence to technoscientists in their efforts to find universal languages and universal explanations for realms of activity that used to be conceptualized as separate and subject to different sets of laws.[4]

The focus of these interdisciplinary OR teams was not one of advancing the frontiers of knowledge; they were not part of the tradition of scientific practices existing within the "narrative" of the search for ultimate truth. Rather, the focus was on solving very specific problems arising out of the war situation—problems which changed often and sometimes unpredictably.

The whole orientation of OR was, and is, toward utility, or "performativity," to use Lyotard's term (Lyotard 1984, 46). Thus, the specific "content" of a discipline is of little importance when compared to the specific techniques of observation and calculation that are developed by scientists that compose the OR groups. In OR work team members are encouraged to abstract mental techniques from their disciplines and apply them to concrete problems of operationality. These mental techniques are augmented with novel quantifying techniques, some of which were refined in OR work. These techniques included statistical summarization and inference, Monte Carlo simulation, queuing theory, and, eventually, game theory.[5]

The focus of Blackett's group was not on the weaponry itself or on the invention of new weapons but on the relation of weapons and personnel and of weapons and specific military tasks or operations. As Blackett himself stated, "relatively too much scientific effort has been expended hitherto on the production of new devices and too little in the proper use of what we got" (Morison 1963, 125). According to Samuel Morison, when scientists were given a problem such as how to advance antisubmarine warfare, they asked themselves such questions as: "What pattern of depth charges at what settings has the best mathematical chance of killing a submarine? What disposition of escorts around a convoy gives optimum protection?" (Morison 1963, 125).

The kind of investigation and questions that OR pursued had a profound impact not only on the specific organization of the research but on the relations between scientists and the military. Blackett's group, for example, had decided early on that its work could not be done at a centralized laboratory; team members had to be in the field. Morse, in his autobiography, reports Blackett as commenting that

> early-warning radar did not become fully effective until some of the civilian laboratory men went to the radar installations to work with the

military operators. These "scientists in the field" helped the military find out what the radar could do under wartime conditions; they themselves also learned at first hand the inadequacies of their first designs when used by non-technical personnel. (Morse 1977, 173)

Blackett ended up fielding groups of scientists with several units of the British military such as the Fighter Command and the Coastal Command, where scientists focused on specific operational problems of those units. Although Blackett did establish a central office, this office functioned as an administrative and training unit, rather than as a center for actual investigation. The end result was the improvement of several aspects of military operations in a very short time. Morse characterized the contributions of Blackett's group in this way:

> Given time enough, the military would, of course, have learned by themselves how the new gear worked and would have been able to improve the man-machine interface. But there hadn't been enough time for this traditional process to succeed. The Battle of Britain had been too urgent to be trusted entirely to the military. (Morse 1977, 173)

In Britain the practices of OR appeared in large part during World War II. In the United States Thomas Edison has been credited with practicing OR during World War I. As chairman of the Naval Consulting Board, Edison proceeded to do statistical research on the sinking of British and French ships in the English Channel area by enemy submarines. He compiled an impressive array of information and used it to infer the best ways of avoiding submarine attacks. A particularly interesting aspect of Edison's work was his creation of a tactical game board with pegs representing ships and submarines. Edison crafted a similar board for an opponent, and they played a war game in which the goal was to find out ways of defeating submarine surveillance of ships. According to William Whitmore, who made a presentation on Edison's work to the OR Society of America, Edison found that, by following certain steps, most ships could safely navigate to their destinations without being spotted by enemy submarines (Whitmore 1953).

Whitmore states that very little came out of this work, even though it was apparently quite sophisticated. He attributes the failure to the fact that the Naval Consulting Board was external to the functionings of the military, and little of its research trickled down to operational commands. The relationship between the military and scientific advisors was to change dramatically during World War II when OR scientists and others would circulate widely within the military itself.

The nonmilitary beginnings of OR practices in the United States can be traced to the work of Horace Levinson, who worked for the Bamberger chain of stores from 1924 until 1946. During that period Levinson, an astronomer who decided to shift his career to business, produced studies that included customers' buying habits and relations of type of merchandise and demographics of neighborhoods surrounding specific stores, as well as studies looking into cash management (Solow 1951; Trefethen 1954). Levinson defined his work as "the application of the scientific method to the study of the operations of large, complex organizations, in order to give executives a quantitative basis for decisions that will increase the effectiveness of such organizations in carrying out their basic purposes" (Solow 1951, 106).

In spite of the early work by Edison for the military during World War I and the successful work of Levinson in the retail business during the 1920s and 1930s, the American military establishment did not pick up OR techniques until after the Battle of Britain. It is necessary to highlight this fact of discontinuity because, even though OR existed before, it was not until World War II that it came into wide use. The war served as catalyst for many new "rationality practices." While historically we can trace the beginnings of OR prior to World War II, the advent of World War II was the most significant factor in the widespread adoption and extension of these techniques, combining it with such technologies as game theory and systems analysis. These distinct but related techniques may be thought of as "dispersed rationalities."[6] During and after World War II we see the combined appearance of similar practices with distinct historical vectors that marks the discontinuities I am trying to make explicit.

In the fall of 1940, James Conant, president of Harvard University, visited England in his capacity as chairman of the National Defense Research Committee (Fortun and Schweber 1993, 603). There he became acquainted with the usefulness of OR to the functioning of the Air Defense System.[7] Soon after the United States entered the war, OR became a fixture of the three branches of the American military.

One of the OR groups set up was the Antisubmarine Warfare Operations Research Group under the aegis of the navy but with reporting duties to the army as well (Fortun and Schweber 1993, 603). This group was headed by Morse of MIT and by William Shockley of the Bell Telephone Laboratories (Baxter 1946). Some scientists in this group were afforded many opportunities for observing combat operations in several theaters of war—in both the Atlantic and the Pacific. The studies of this group involved "the range of detection by eye, by radar, or by sonar gear, and, more important, the search rate, or number of square miles which a given craft can search in an hour" (Baxter 1946, 405). Again, the research of OR was tied to specific combat missions. The intimate contact with the urgency of a given challenge has given OR and related practices a clear orientation toward performativity—a characteristic of many forms of knowledge emerging out of World War II. The emphasis on solving "real world" problems rather than "ivory-tower" ones was also a feature distinguishing the work at the Servomechanisms Laboratory, and it has been a trademark of Jay Forrester's work and the work of others trained in System Dynamics.

According to Baxter, one of the immediate results of OR work based on actual combat operations was the realization

> that this work could be best described in terms of probability rather than a definite range of detection. From this probability of sighting could be computed the average range under certain conditions and the effective search rate. Operational data was punched on IBM cards and then analyzed by machine methods. On the basis of a sighting probability curve, obtained from operational data, different aircraft search plans could be compared in efficiency, and the best plan found. (Baxter 1946, 406)

The plans that were developed from this work became the official operating doctrine of the antisubmarine units of the navy. According to Baxter similar studies were done for radar and sonar search plans as well as for radio sonobuoy searches.

One interesting aspect of the work of the Morse group was that they had a fair amount of mobility within the military. This mobility was not without setbacks or resistances, but it was there nonetheless. In his autobiography Morse talks about his early days working in a military context as being in an "alien environment" (Morse 1977, 76). This feeling was compounded by a certain ambiguity about the status of group members. They wondered, "were we the equivalent of officers or of ordinary seamen?" (Morse 1977, 176). Eventually, the scientists turned this ambiguity into an advantage by talking as equals with both officers and regular seamen. Their ambiguous status allowed them to circulate with a fair amount of freedom within the different units they served. This was true on both sides of the Atlantic. Such freedom was not without problems, however. Morse narrates the case of a staff member of his group who, having grown impatient with the need to show his identification card, was shot and wounded by an overzealous guard. Morse's wry comment was that "the group members were intelligent people; they learned fast" (Morse 1977, 159).

A New Regime of Machine Vision

I have discussed above the characteristics of OR, some studies done using OR, and its growth during World War II. Now I turn to the relationship between OR and a new regime of vision, one characterized by a shift from reliance on the human eye as the center of vision to one relying on machinic vision.

The quotation from Steinhardt at the beginning of the chapter includes the phrase "geometrical wanderings." Steinhardt uses the phrase to describe the type of movement that ships and planes would engage in once their patrolling moves were planned with the aid of OR calculations. These mathematically precise wanderings sharply increased the ability of the antisubmarine surveillance to spot U-boats. They also point to a radical new way of "seeing." They displaced, at least as far as these combat operations were concerned, the roving, generally random observation of the human eye onto a form of seeing that had preestablished patterns of observation prior to the human eye's doing any actual observation. The conjunction of OR with the needs of surveillance served as the basis for a new theory and practice of sighting. Morse and George Kimball talk about a new theory of sighting in relation to submarine surveillance. They state:

> Before the operational data on the visual sightings of submarines by aircraft were completely understood, it was necessary to develop a new theory of sighting, which required considerable mathematical analysis and also a certain amount of physiological and psychological experimental work. (Morse and Kimball 1951, 6)

One of the interesting characteristics of the new way of sighting is the impetus toward automation of vision and its connection with specific forms of mathematization of space, in this case the oceans.

Dependence on machine sighting developed concurrently with new objects of observation made possible by such new technologies as radar. Eventually radar, OR, and automation of vision came together in such specific weapon systems as antiaircraft artillery. New techniques of seeing included a surveillance of invisible objects, either because they lurk underwater or beyond the horizon of the human eye. Out of the complex array of radar, OR, and cathode ray displays, a new regime of vision took shape. This new regime did not focus on the detailed specificity that the human eye is educated to see, but rather on the electronic, sonic, infrared, and ultraviolet traces and shapes that organisms and machines have, produce, and leave in their wake. What is seen is often the probabilistic aggregated traces. What becomes ultimately important to see, because it is an issue of life or death in combat situations, is precisely that which the human eye can no longer see. It is here we find the emergence of the new regime of vision, dependent on machines for seeing, which Paul Virilio calls "the eyeless vision" (Virilio 1989, 2).

This new regime of vision has grown enormously in the last twenty years or so and is at the core of technologies such as computer simulation and virtual reality. Its extensive presence was certainly seen and felt during the Gulf War, to some extent a virtual war dependent on machinic vision, such as satellite surveillance, for its conduct.

The new regime of vision that emerged out of World War II is also central to the emergence of global spaces. These spaces cannot be encompassed with the naked eye. Unlike the spaces of nation-states, with their borders marked by physical elements such as rivers and mountains, we cannot see the borders of global spaces. We see global spaces, largely conceptually, when we think with the use of computer simulation, for example. We see the earth in its totality only from outer space, with satellite vision. The confluence of satellite vision and computer simulation allows us to grasp a sense of global spaces that is not immediately describable by what the eye sees. The spatial sense of the global earth is achieved, then, largely through mental models that are dependent on computer modeling and simulation and the machinic vision of satellite imaging.

Systems Thinking and the Blurring of Boundaries between the Organic and the Technological

As noted above, the emergence of discourses and practices of globality is dependent on a set of technologies both material and mental. Global spaces do not exist independently of the set of technologies that make them possible. As knowledge is produced about global spaces, these spaces achieve their concreteness and materiality. In other words, global spaces are not a priori ahistorical essences waiting to be discovered, and they are not the figment of someone's imagination. Rather, they are scientifically, politically, technologically, and culturally constructed, in large part through the hard work of knowledge production.

Important to discourses of globality is the encompassing field of intervention they make possible. This field of intervention includes both organic and machinic elements that in older explanatory modes could not be treated similarly because they were perceived as responding to incommensurable sets of laws. Thus, machinic processes of production are now conceptualized as belonging to the same field as natural processes. For example, the article "Strategies for Manufacturing" by Robert Frosch and Nicholas Gallopoulos in the *Scientific American* issue discussed in the first chapter argues for the need to think of industrial production moving toward a model they call "industrial eco-system," characterized by decreasing use of materials and closed-system manufacturing (Frosch and Gallopoulos 1989, 144). The discussion of Forrester's modeling in the next two chapters is another example of how the machinic and the organic are now conceived as susceptible of explanation within the same theoretical field.

Central to the erasing of boundaries between conceptions of the natural and the artificial was the emergence of new objects of study, objects that arose out of the very practices that were invented in order to look at them. The practices that made it possible to conceive of new objects of research were in fact largely constitutive of those very objects. Many of these practices and their new objects were conceptualized with and through the language of systems thinking, feedback loop mechanisms, and cybernetics.

Historian Donna Haraway, for example, examined the emergence of these new objects in the scientific practices of E. O. Wilson, the distinguished entomologist and controversial proponent of sociobiology who started his career in World War II. Haraway argues that the new objects of study

> can only be constituted in the basis of material practice composed of the daily elements of scientific work; field and laboratory custom, machinery, social hierarchies and networks, funding possibilities, contests for privileged model systems, metaphor, struggles over language. (Haraway 1981–82, 245)

In her article Haraway explores the rise of sociobiology, or more specifically "its theory of the machine-organism as command-control-communication system," in the context of wartime science. According to her, E. O. Wilson's theory posited new "machine-organisms" structured as "problems in information exchange for production control and military strategy, both cast in terms of replicating cybernetic communications systems" (Haraway 1981–82, 245). It is important to note that Haraway does not maintain that these new "technical-natural" objects and the related practices and discourses were uncontested. This was certainly not the case. Rather, they became hegemonic, but they never completely displaced other modes of organizing knowledge.

Haraway looks at the work of E. O. Wilson as one example of how wartime science and the material scientific practices emerging from that period constituted "natural-technical objects of study as a military-industrial command-control system" (Haraway 1981–82, 245). In particular she focuses on Wilson's early work on the chemical communication of fire ants. Wilson establishes a taxonomy of messages within the ant colony and proceeds to study the colony through an analysis of the message structure. As Haraway states, Wilson aimed at the "dis-assembly and re-synthesis of insect society through the identification and manipulation of its constitutive communication elements" (Haraway 1981–82, 257).

Within this cybernetic framework Wilson proceeded to study the foraging behavior of the ants. He established the relation between a food source and the ant colony as a complex feedback system of communications. The number of ants would build up at an exponential rate at the food source, until the negative feedback of some ants unable to reach the food would restore the initial "overshoot." Haraway points out that this description is not mere metaphor but a causal explanation of how ants reach food sources and communicate, through the negative feedback of returning ants, the quantity of food. Wilson's study is quite complex and continues by exploring the food foraging of ants with an artillery analysis model of accuracy vis-à-vis the food sources. What Haraway finds interesting in Wilson's explanatory scheme is his insistence on a "probabilistic analysis, the dissection of society into patterned communication events, and the adoption of electronic metaphor-concepts like negative feedback to suggest causal analysis" (Haraway 1981–82, 257).

Haraway's article is a brilliant discussion of the ways in which scientific practices bring out new objects of analysis, like the cybernetic system composed of the fire ants, food sources, and the chemical communication within the ant colony. In Wilson's explanatory model what to someone like Vernadsky would appear as distinct natural objects such as ant colonies and food sources gets transformed into a cybernetic system that blurs the former boundaries among natural objects and between natural and technical objects.

These types of cybernetic conceptualization carry over into other systems, including those where human beings play a part. In ergonomics, for example, humans are conceived of as "receiver-operators" within a larger communications system (Haraway 1981–82, 250). It is important to remember, as Haraway makes clear, that ergonomic science shares the same conceptual roots as E. O. Wilson's ideas, all emerging from the experiences of World War II.

It should be said that the migration of the military language of command and control to the areas of biology and ecology is not a one-way movement. Language from biology also found its way into military descriptions. For example, a 1950 report on the Air Defense System, quoted by the historian Evelyn Fox Keller, states that the air force system should be thought of as an organism.[8]

In the field of ecology, the modes of conceptualization emerging after the war were also profoundly influenced by cybernetics and systems thinking. Evelyn Hutchinson, an ecologist at Yale University, was a pivotal figure in the development of ecological notions after World War II. He was instrumental in introducing Vernadsky's ideas to an American audience. He taught Robert MacArthur, a close associate of E. O. Wilson in his early years as a scientist, and he was the teacher of H. T. Odum, who developed the systems ecology that dominated the ecological field in the 1960s and the 1970s.

Hutchinson's interest in cybernetics and its relation to ecology occurred when he attended a Macy conference in 1946. These inter-disciplinary conferences, of which over 130 were organized within a ten-year period, were sponsored by the Macy Foundation in New York City. Their main purpose was to advance the perspective of complex systems as self-regulating feedback systems within many social and biological fields (Heims 1991). These gatherings were quite influential because they per-mitted an interdisciplinary cross-pollination between individual scientists from a diverse array of disciplines. From these conferences scientists wrested concepts from other disciplines, research practices and models which could be adapted in their respective practices. The aggressive interdisciplinarity of the conferences and the aura of optimism added to their long-term influence.

The title of the paper Hutchinson presented at the cybernetics conference was "Circular Causal Systems." The essential conclusion of the paper was, as ecologist Peter J. Taylor writes, "groups of organisms are systems having feedback loops that ensure self-regulation and persis-tence" (Taylor 1988, 217). Hutchinson's paper became another example of a growing number of works that proposed a reconceptualization of nature as a system in which the old organismic tradition with sharp separations between the organic and the inorganic was steadily erased. In the new systems view, states Taylor, "living and non living feedback systems alike obeyed common mechanical principles, including their mode of evolution" (Taylor 1988, 221).

Hutchinson's ideas were further elaborated by his student Odum who was strongly influenced by cybernetic principles. In his doctoral thesis Odum described ecology, quoting Norbert Wiener's definition of cybernetics, as "one part of the study of mechanisms of steady states in all types of systems" (Taylor 1988, 225). Odum eventually included human beings in his description of ecological systems. As he put it, "the old systems and the new are being joined into an overall network including factories and towns, reefs and grass flats, and the flows between them" (Taylor 1988, 231).

Odum's work is clearly a product of the new forms of knowledge that emerged out of World War II. His ideas were largely influenced by the models that sprang from wartime research in areas such as automated aiming and firing mechanisms, with their heavy emphasis on feedback loop systems theory. Even the funding for his research was underwritten by institutions such as the Office of Naval Research (ONR), the National Science Foundation, and the Atomic Energy Commission, all of which were conceived for the funding of science during or immediately after the war. It may not be surprising that Odum eventually turned his gaze to war itself as something amenable to systems analysis study. In his textbook *Systems Ecology: An Introduction*, published in 1982, Odum includes the illustrations of two systems models, one depicting the effect of tides and hurricanes on nutrients flows of a mangrove swamp and the other the effect of flows of money into the Vietnam War. For the Vietnam model Odum helpfully lets his reader know that "B52s and so on were converted into energy at the rate of 14,000 calories per dollar" (Taylor 1988, 243).

The discourses and practices centered on OR systematized and normalized the existence of interdisciplinary groups of knowledge production which spanned many disciplines and which also crossed the government/private and civil/military divides. Operations Research, especially when used in combination with tools such as radar, also advanced the migration of mental and sense functions from the human body to external machines, as illustrated by machines endowed with vision-sensing equipment.

Systems thinking advanced modes of representation that permitted things to be seen as being interconnected in larger wholes. The emphasis of this mode of looking at the world is focused less on individual performance than on the functioning of the system as a whole. The focus on the system as a whole, in turn, necessarily weakens or blurs the boundaries between nature and technology. The language of feedback loops, circuits, and flows is applied to both organic and technological entities, thus doing away with the distinctions between the two realms.

The relations between OR, systems thinking, cybernetics, game theory, and information science, together with the use of computers and radar, can be seen as the core of a formidable technoscientific apparatus with which to understand and manage the world.[9]

In the first chapter I stated that the last twenty years or so have reg-istered an increasing number of discourses, such as studies and reports, focusing on the planet as a whole. I also showed how these discourses depend on specific practices, such as computer modeling and satellite imaging, to gain support and legitimacy. I further demonstrated that the aggregation of these discourses and practices brings to the forefront a concrete reality that looks new and strange when seen from the perspec-tive of the modern era now disappearing. Many of the distinctions and boundaries that undergirded modern knowledge become less distinct in this new world. Thus, the relations between the natural/organic realm and the technological realm are blurred and encompassed in the same set of explanatory laws. Systems thinking is at the core of this new explanatory language. I also suggested that the relations between humans (or "man," to use the historically accurate term emerging from the Enlightenment) and nature have undergone a radical restructuring. Humanity goes from being the overlord of nature to being yet one more species in the bio-sphere and in potential danger of sufficiently changing the planetary conditions of life to engineer its own extinction.

In the second chapter I have suggested that knowledge production emerging from the maelstrom of World War II had a profound impact in undermining the solid distinctions that characterized the modern era. Thus, out of the day-to-day exigencies of the war and the need to defeat a cunning and brutal enemy emerged new ways of thinking, such as OR and systems thinking, that proved invaluable in the war effort. Out of that success these new types of thinking, which I call "dispersed rationalities," became increasingly embodied in broader and broader areas of knowledge, such as management, biology, and ecology. In due time they transformed knowledge. The world we live in today, where the whole planet can be thought of as a system amenable to management, is profoundly dependent on these knowledge transformations.

In the following chapters I will look in detail at a particular exemplar of how these knowledges went from the localized experiences of the war to the space of the whole planet. I will also explore how these knowledges were transformed as they encountered new problems and concerns, such as the growing effects of human production on the planet as a whole.

Notes

1. From his article "The Year 2000 Will Not Happen" (Baudrillard 1986). The whole quotation reads: "We leave history to enter simulation (as much, to my mind by the biological concept of the genetic code as by the media, as much by space exploration, which for us functions as a space of simulation, as by the idea of the computer as a cerebral equivalent, etc.). This is by no means a despairing hypothesis, unless we regard simulation as a higher form of alienation—which I certainly do not. It is precisely in history that we are alienated, and if we leave history we also leave alienation (not without nostalgia, it must be said, for that good old drama of subject and object)" (Baudrillard 1986, 23).

2. I have taken the notion of extending and modeling of scientific practices from Andrew Pickering. See, for example, Pickering and Stephanides (1992).

3. The heterogeneity of OR teams is discussed in Fortun and Schweber (1993, 607).

4. A point that Geoff Bowker (1993) and Steve Heims (1991) also make.

5. See Rider (1992).

6. I borrow the use of the term "rationalities" from Andrew Pickering to describe these techniques. See generally Pickering (1995). I have added the descriptor "dispersed" to indicate both the distinct historical and sometimes geographic vectors that characterize the appearance of these rationalities and the fact that they permit the dispersion of functions across a managerial field facilitating the decentralization of management and surveillance functions.

7. The relations between the Americans and the British were not without difficulties. Conant was generally suspicious of some of the intentions of the British and decried their colonial policies (Hershberg 1993).

8. The report, entitled "Progress Report of the Air Defense Systems Engineering Committee," is quoted by Keller in her book *Refiguring Life* (1995). The report states, in part, that there are three kinds of organisms: "animate organisms which comprise animals and groups of animals, including men; partly animate organisms which involve animals together with inanimate devices such as the ADS (Air Defense System); and inanimate organisms such as vending machines. All these organisms possess in common: sensory components, communication facilities, data analyzing devices, centers of judgment, directors of action, and effectors, or executing agencies" (Keller 1995, 90).

9. Keller sees very important commonalities between the disciplines, such as OR and systems thinking, so that she uses the term "cyberscience" to refer to all of them. She points out that all of them share the common goal of analyzing complex systems. They also share a common conceptual language and similar mode of representation (Keller 1995, 84). Similarly, Pickering (1995) refers to the assemblage of knowledges emerging from World War II, together with their many institutional forms, as "cyborg regimes," a term that, like "cyberscience," neatly captures the commonality of these forms of knowledge production.

Three

From Servomechanisms to Planet Management: Jay Forrester and System Dynamics

In the opening chapter of his first book, *Industrial Dynamics* (1961), Jay Forrester identifies what he understands as the four pillars or foundations on which better understandings of social organizations are possible. Forrester lists, as by-products of military systems research, "the theory of information-feedback systems; a knowledge of decision-making processes; the experimental model approach to complex systems; and the digital computer as a means to simulate realistic mathematical models" (1961, 14). This chapter is an exploration of how Forrester came to see the four above-mentioned foundations as constitutive of the type of management and thinking tools that he developed. A clear historical analysis of this process is important to our story because it clearly shows how the new techniques for apprehending and thinking about the world that were developed during World War II subsequently became embedded in the discourses and practices that have facilitated the conceptualization of globality.

The 1940–41 academic year saw the arrival in Cambridge, Massachusetts, of a bright farm boy from Nebraska. Jay Forrester, having completed a B.A. in engineering at the University of Nebraska, had been admitted to the masters program in electrical engineering at MIT. Forrester reports that he chose MIT over other possibilities for two reasons:

> First, they offered me a $100.00 per month research assistantship, which was more money than any other university had offered. Second, my mother, from her library experience in Springfield, Massachusetts, knew about MIT. In the high plains of the United States at the time "MIT" more

often implied salesmen for a financial institution, the Massachusetts Investors Trust, than an engineering school. (Forrester 1991, 7)

It was a choice that would have a profound effect on Forrester's ability to do engineering work and to eventually contribute in several important areas, such as the development of the digital computer, the SAGE (Semiautomatic Ground Environment) defense system, and management studies.

During his first semester at MIT, Forrester worked part-time as a research assistant in the High Voltage Laboratory under the direction of John Trump. In the second semester of that year, he was recruited, or as Forrester jokingly put it, "commandeered" by Gordon Brown, an engineer then in charge of the newly created Servomechanisms Laboratory.[1] It was a recruitment that would also have a deep impact on Forrester and the future direction of his research and career. Forrester and Brown developed a deep mentoring relation and friendship that has continued to this day.[2]

The Servomechanisms Laboratory

The Servomechanisms Laboratory was born out of the rumblings of war in Europe in 1939 and the realization by the U.S. Navy of a need to upgrade the rapid-gun positioning equipment aboard their ships (Wildes and Lindgren 1985). To that end the navy asked MIT to train a group of officers in the latest technologies of servomechanisms and fire control. The year-long course was taught by Brown, who eventually turned the notes for the class into a textbook (Brown and Campbell 1948).

While working on improving the servo systems used to control large guns, Brown developed an increasingly broader relationship with the Sperry Gyroscope Company, which was at that time working on an antiaircraft sight for large thirty-seven-millimeter guns (Burchard 1948). This relationship developed into an agreement between Sperry and MIT to design servomechanisms for the navy. The advent of World War II broadened the activities of the laboratory and its staff, which at the height of the war numbered over 100 members. The laboratory developed several remote control devices, using servomechanisms, to drive large gun turrets, radar equipment, and ship antennae (Noble 1984). It was at the beginning of this growth period for the laboratory that Brown recruited Forrester.

While working in Brown's laboratory, Forrester was involved in developing servomechanisms to control gun mounts and radar antennae

(Forrester 1991). His work focused on designing hydraulic mechanisms—
a bit of a departure for Forrester whose training was in electrical engineering. The biggest reason, according to Forrester, for a concentration
on hydraulic servomechanisms was the army's distrust of electronics on
anything except radios.[3] The work that Forrester was involved with was
oriented toward

> an extremely practical goal that ran from mathematical theory of control
> and stability to the military operating field, and I do mean the operating
> field. At one stage, we built experimental hydraulic controls for a radar
> designed at the MIT Radiation Laboratory. After redesign, the radar
> was intended for aircraft carriers to direct fighter planes against enemy
> targets. The captain of the carrier U.S.S. Lexington visited MIT and
> saw this experimental unit, which was planned for production a year or
> so later. He said, "I want that; I mean that very one; we can't wait for
> production equipment." He got it. (Forrester 1991, 8)

That particular unit containing the experimental hydraulic controls was in operation for nine months before malfunctioning. Forrester
went to Pearl Harbor to repair it, but the ship could not wait for him
to finish the repairs so the USS *Lexington* sailed to war with Forrester
aboard. Forrester ended up involved in the invasion of Tarawa island
off the Marshall Islands. Eventually the USS *Lexington* was disabled by
enemy fire and returned to port. Forrester's wry comment on his actual
involvement in combat situations while repairing the machines was that
"the experience gave a very concentrated immersion in how research and
theory are related to practical end uses" (Forrester 1991, 8).

It is worth highlighting the intense focus on the practical application of Forrester's work. Even when involving theoretical considerations,
the ultimate goal was solving a concrete problem. This practical orientation also characterized OR and was the ethos of the Servomechanisms
Laboratory under Brown. Many of the technoscientific practices that
came out of the war experience were informed by a strong ethos of
performativity and utility. Forrester's System Dynamics was no exception.

After the war, Forrester considered starting a company dealing with
feedback control systems. However, after talking things over with Brown,
he decided to take on a project that Brown offered him—that of building
an aircraft flight simulator (Forrester 1991). In 1944 the navy and MIT
had agreed to collaborate to build an Airplane Simulator and Control
Analyzer (ASCA), with the navy providing the funding and MIT supplying
technical knowledge.

The immediate goal of the ASCA program was to build a "dual-purpose flight simulator" (Redmond and Smith 1980, 1). The purpose of such a machine was to train crews in the use of fighter planes. The main object of a flight simulator is to mimic on the ground the behavior of a flying aircraft. This objective was achieved by "the creation of a dynamic representation of the behavior of an aircraft in a manner which allows the human operator to interact with the simulation as part of the simulation" (Rolfe and Staples 1986, 3). This dynamic representation was originally obtained through the use of pneumatic controls, which were eventually replaced with electrical motors (Redmond and Smith 1980).

In the case of ASCA, the original goal was to have a machine capable of (*a*) training new pilots and (*b*) calculating the flying characteristics of planes under development. The latter capability, which was conceived of by Admiral Luis de Florez in consultation with MIT engineers, was to be achieved by combining the data generated by the interactions of a human operator (i.e., pilot-in-training) and the simulator with data generated by the simulator within a wind tunnel.[4] The ultimate goal was to collect the data generated by the human operator in the process of responding to the behavior of a new or untried plane design (Redmond and Smith 1980).

The ASCA machine was never developed.[5] Instead, the project, after many bureaucratic and technical twists and turns, evolved into Whirlwind, a project designed to develop a digital computer.[6] Intrinsic to ASCA was the call to build a "specialized calculating machine that could be configured for a particular airplane in response to data obtained by experimental means, and the pilot's control motions could be fed into the system by actually having the pilot fly the resulting airplane" (Redmond and Smith 1980, 4).[7] The original "calculating machine" was an analog computer, which proved insufficient for the task of processing a great deal of information sufficiently fast to approach "real time."

At that point in his capacity as director of the ASCA project, Forrester had a conversation with Perry Crawford, an MIT graduate working with de Florez in the Special Devices Division, who encouraged Forrester to explore the field of digital computers, then in its infancy, as a way to overcome the limitations of analog computing (Forrester 1991).

Three aspects of the complex and involved Whirlwind project need some mention. The first one is the conceptualization of the Whirlwind computer as a "universal computer." In an early 1946 letter to Lieutenant Commander H. C. Knutson of the Special Devices Division, Forrester outlined his understanding of what he called a "Universal Computer." Forrester predicted that a digital computer, such as the one he was designing, would have broad military and civilian uses. According to historians Kent Redmond and Thomas Smith, Forrester predicted a

Universal Computer with definite possibilities for military application in both tactics and research. In tactical use it would replace the analog computer then used in offensive and defensive fire control systems, and furthermore, it would make possible a coordinated CIC (Combat Information Center) possessing automatic defensive capabilities, an essential factor in rocket and guided missile warfare. In military research, electronic computation held promise of possible wide and diversified research program in "dynamic-systems:" 1) aircraft stability and control; 2) automatic radar tracking and fire control; 3) stability and trajectories of guided missiles; 4) study of aerial and submarine torpedoes, including launching characteristics; 5) servomechanisms systems; and 6) stability and control characteristics of surface ships. (Redmond and Smith 1980, 42)

In his letter to Knutson, Forrester also predicted many civilian uses in the natural and social sciences for a digital computer. In his list he included nuclear physics, thermodynamics, mechanical and civil engineering, and statistical studies. He concluded his letter with the following statement:

The development of electronic digital computation is only beginning, and considerable effort and money will be expended in achieving the equipment to meet the above objectives. Once sufficient development is completed, however, the cost of duplicating electronic computing equipment will be less that for other forms of computers. Beginning with a suitable basic design, new computers could be built with facilities for a specific magnitude of problem by adding or omitting standardized memory or storage units without requiring significant redesign. (Redmond and Smith 1980, 42)

What is especially significant in Forrester's statements is his insistence on the universality of computer applications. Certainly, the letter was part of the process of securing funding, a problem that plagued the Whirlwind project for a long time, and so a certain amount of hyperbole is to be expected. Nevertheless, the statement is an important example of the ways in which discourses about computer uses came to be. These discourses became an intrinsic component of System Dynamics and of many other related managerial and calculative tools. The specificity of the list that Forrester produced in 1946 suggests a great deal of confidence in the potential of the newly developing technology. It also suggests a particular way of looking at the world—a way that assumed the computerized simulation of processes both possible and necessary.

Paramount in these simulations is the study of military "dynamic systems." Ten years later, the simulation of dynamic systems was extended by Forrester to include nonmilitary ones, such as industrial processes and their management. Another significant aspect of Forrester's letter is his comment that the technology of "universal computers" would become easily replicable and thus advance their very universality. While this is an idea taken for granted today, it was not widely circulated in 1946. In fact, most computers, whether analog or digital, were built for specific and fairly narrow purposes and thus perceived to be connected to those specific purposes and sites. Replicability of technology, where computers become one of a series, greatly aided in making computers into "universal" machines. Thus, the impetus to universality embodied in mental technologies, such as cybernetics, finds its counterpart in the material technology of the digital computer as imagined by Forrester and embodied in the Whirlwind digital computer (Bowker 1993).

A second important outcome of the Whirlwind project was Forrester's invention of a new type of computer memory and his increasing experience in managing large projects. The problem of appropriate memory for the increasing computing demands that the Whirlwind project envisioned was becoming something of a bottleneck in developing digital computing. It was a problem to which Forrester dedicated considerable thought. By 1949, he was spending a fair amount of time working on alternative methods of information storage. On June 13 of that year, Forrester made an entry in his engineering book detailing a system that would eventually be known as coincident-current random-access magnetic computer memory, or RAM for short (Pugh 1984). It was an invention that eventually brought Forrester a degree of personal fortune and which, in the words of historian Emerson Pugh, "substantially altered the evolution of stored-program computers and, indeed, the entire computer industry" (Pugh 1984, 62). It was an invention characteristic of the results-oriented work that Forrester engaged in throughout wartime and the early postwar era. It illustrates his facility in moving from theoretical questions in control theory to solving vexing practical application problems at the level of machine performance. This ability, which he shared with other engineers educated during World War II, partly explains the ease with which he traverses the machinic, social, and managerial divides in his System Dynamics. For him there is a continuum and an underlying set of approaches that should work equally well in what, to others, appeared as incommensurable realms.

It should be added that RAM was not the only invention to emerge out of the Whirlwind project. As historian Bernard Cohen comments:

The new technologies proposed and tried for Whirlwind are like a roster of the cutting edge of the computer art—they include storage of data, new types of circuits, the use of magnetic tape, transistor circuitry, numerical methods, checking devices, computer reliability, and computer programming. By 1951, high-speed techniques developed for Whirlwind were making their way into air-traffic control, industrial process control, insurance handling, inventory control, economic analysis, and scientific and engineering computations. (Cohen 1988, 141)

As director of the Whirlwind project, Forrester also gained increasing managerial experience. The Whirlwind team employed twenty-five engineers plus an equal number of support personnel (Pugh 1984). And, while Forrester had excellent managerial help from Robert R. Everett, he was very much in control of the whole administrative process in addition to being the leading engineer and researcher in the program.[8] Besides gaining managerial experience with very large systems of people and machines, he learned how to work with large bureaucracies, such as the military. Forrester offered a humorous, and telling, example of what this involved. He recalls:

> The Admiral [Luis de Florez] taught me a number of helpful insights about dealing with government bureaucracies. Later, when we were building a digital computer, we needed another hundred thousand dollars to continue. The response from de Florez was "impossible! That is above my approval authority and too little to justify going to the Secretary of the Navy. You must ask for either fifty thousand or two hundred thousand." We chose the latter figure. (Forrester 1991, 9)

Forrester's ever-increasing skill as manager became even greater with his involvement in SAGE, a land-based air defense system.

Before discussing Forrester's involvement with SAGE, we need to consider, at least briefly, another important step in Forrester's career—namely, his connection to a program initiated in the Servomechanisms Laboratory to produce what became known as "numerically controlled machine tools" (Noble 1984, 84). The project was started in 1949 when John Parsons of the Parsons Corporation of Traverse City, Michigan, contacted Gordon Brown at MIT. The Parsons Corporation, at the time, was an important manufacturer of helicopter rotor blades for the military. Brown offers this narration of the contact:

> It so happened that John Parsons telephoned me one afternoon from Worcester to ask if I could furnish him with a power drive that would take

intermittent data. After a little discussion of what he meant by power drive, it seems that he had in mind something of a couple of horsepower, and what he meant by intermittent data turned out later to be pulses that would come from some punched cards, or the like, that would carry the instructions for the control of a machine tool. I told him that about $25,000 was a round figure for a program of that kind, and that was the beginning of the program which became Numerical Control.[9]

What had originally attracted Forrester to this project was the possibility of exploring the use of digitally controlled servomechanisms. As David Noble points out:

> MIT had considerable experience in analog servomechanism control but had only just begun to contemplate digitally controlled servo system. The ASCA would have been such a system, but it was abandoned when the decision was made to concentrate exclusively upon the computer itself, Whirlwind. (Noble 1984, 118)

While Forrester was not greatly involved in this project, his participation in it gives us insight into his way of thinking. The problem that Parsons had was, for Forrester, not the main issue but rather was an avenue to explore the wider use of digitally controlled mechanisms. The possibility of extending digital technology was much more central to Forrester's interest. While this attitude may be understandable in the context of the cozy relations between MIT and the military, where researchers had some latitude to explore beyond the immediate problem at hand, it was to be an expensive lesson for the Parsons Corporation on the limits of partnering with a research university. The Numerical Control program, in fact, soon became much broader than initially anticipated by the Parsons Corporation. Ultimately, Parsons dropped out of the project feeling that MIT had abused their relation by expending much more money than anticipated and by taking some of the ideas that Parsons had brought to the table and claiming them as its own.[10]

Financial Problems of the Whirlwind Project

While the Servomechanism Laboratory and Forrester, to some extent, were involved with the numerically controlled project, the financial problems of the Whirlwind project, in the form of cost overruns, were by 1949 threatening to terminate the program. Some members of the

ONR, Mina Rees in particular, were very concerned about the costs of the program. The ONR compared Forrester's project to a similar digital computer program at Princeton University, directed by John von Neumann, and found that the Forrester project was much more expensive (Wildes and Lindgren 1985). At one point Rees called the Whirlwind project "unsound on the mathematical side" and "grossly over-complicated technically" (Noble 1984, 112). Historians Redmond and Smith attribute the problems between mathematicians at ONR, such as Rees, and Forrester to their different conceptions of what Whirlwind was to become. They write:

> If the Whirlwind engineers had not been operating within the protective womb of MIT, it is altogether conceivable that the project would have been terminated by the Navy, particularly after ONR had assumed primary responsibility for Navy research and development. The mathematicians of ONR, enthusiastic about the computer as a scientific instrument of rapid calculation, failed to recognize its potential as a command and control center as early advocated by Forrester and Crawford. . . . The engineers of Project Whirlwind and SDC, concerned primarily with application to military needs rather than development of theoretical concepts, saw it as an instrument primarily adapted to facilitate human control of events in the physical world and only secondarily intended as a mathematician's tool. (Redmond and Smith 1980, 69)

Many years later, in an article reflecting on ONR's computing programs, Rees is more generous in her comments about Forrester and the Whirlwind project. She traces the main differences of view to a perceived ambivalence built into the origins of the program as a simulator. She comments that there was "some confusion as to whether Whirlwind was really a simulator or a general-purpose computer" (Rees 1982, 114). The crux of the issue was the capability of the computer to operate in "real time." The real-time demand was inherited from the ASCA simulator program, when the goal was to incorporate the reactions of the trainee into the training exercise in real time, so the simulator would be felt by the pilot as responding to his actions. Forrester preserved this goal of real time even when the ASCA program was transformed into a full-fledged digital computer program. Mathematicians at ONR were not particularly interested in this feature.

Ironically, it was the real-time orientation that both threatened the program and proved to be its salvation. The explosion of an atomic device by the Soviet Union and the invasion of South Korea by North Korea in 1950 accelerated plans for a continental defense system. At the heart of such a system was the need for a computer that would coordinate and control information from radar stations, guide airplanes to targets, and perform other such functions. Whirlwind became that computer. The air force stepped in with financing, and the project continued (Redmond and Smith 1980). As Rees graciously recalls, "it became clear that we did, in fact, need a computer that operated in real time. Forrester was vindicated, and ONR was saved from having to solve Whirlwind's financial problems" (Rees 1982, 114).

SAGE was an immense undertaking. Run out of the Lincoln Laboratory, it involved setting up a command-and-control defense system. It was described by an air force colonel as "a servomechanism spread over an area comparable to the whole American continent" (Noble 1984, 52). Forrester was made director of Division 6 of the Lincoln Laboratory, which was in charge of designing the real-time computers needed for the SAGE air defense system (Forrester 1991). The problem involved setting up these computers in SAGE computer centers all over the North American continent. The first computers were installed in the late 1950s, and the last one was not decommissioned until 1983 (Forrester 1991).[11]

From SAGE, Forrester learned to have confidence in his notion that real-time computers were an essential tool to manage and control large operations. The conception of real time in Forrester's work was eventually embodied in his System Dynamics, with its insistence on the ongoing *dynamics* of the system under consideration. His modeling, from the earliest version to the latest, involves a conception of real-time dynamics. This process is, of course, made possible with the advent of computers, such as the one developed within the Whirlwind project.

Forrester's vast wartime and postwar experience was eventually channeled into a different activity, that of management education. His war experience became embodied in his management theories and the technoscientific practices he develops to model the endogenous or internal structures of systems.

The Sloan School of Management and the Emergence of System Dynamics

Industrial management is a field of rapidly growing importance in a society with many industrial enterprises presenting problems of a complex technological and human relationship nature. The Foundation and the Institute agree that there is an urgent need for a center of research and education which would embrace an extensive program of scientific research into the problems connected with the management of industrial enterprises and an academic organization for dealing with the subject at both the undergraduate and graduate level. (Killian 1985, 197)

Those words are part of the Deed of Gift and Agreement between MIT and the Sloan Foundation to set up a school of industrial management at MIT with initial funding by the foundation.[12] The Sloan School of Management officially started in 1952, has now gained an international reputation, and is considered one of the top ten business schools in the United States. It still remains notable for its strong emphasis on extending engineering concepts and problem-solving techniques into the field of management.

Forrester became associated with the Sloan School in 1956, just as his work on the SAGE system was coming to an end. This is how Forrester narrates his move to Sloan:

James R. Killian, Jr., who was then president of MIT, brought a group of visiting dignitaries to the Lincoln Laboratory. While we were walking down the hall, Killian told me of the new management school that MIT was starting, and suggested that I consider joining. Discussions over the next several months with Associate Dean Eli Shapiro and Professor Edward L. Bowles led to my becoming a full professor in management. It was my first academic appointment; earlier work had all been on the MIT research staff. (Forrester 1991, 12)

The turn in career directions described above is characteristic of Forrester's propensity to take advantage of what he calls "interesting open doors when they are offered."[13] He has been willing to make fairly large shifts from one field to another. He took advantage of the opening at the Servomechanisms Laboratory when Gordon Brown offered him the job. He then shifted to the development of digital computing (the Whirlwind project and the Numerically Controlled Machines project). He took the opportunity to become the head of the biggest division within the Lincoln Laboratory, in connection with SAGE, as Whirlwind became transformed into a computerized defense system. From SAGE he went to the Sloan School. As discussed below, a series of contingencies led him to apply his System Dynamics first to corporations then to cities and finally to the whole planet. Forrester took advantage of opportunities as they opened, even if doing so risked leaving familiar territory. This propensity was, to some extent, born of the war experiences, where the ability to develop ad hoc solutions and turn to new problems on a dime was highly valued and needed.

While the change from the field of digital computing to management appears abrupt, it was not so in Forrester's case. A large part of his work at SAGE was managerial. As he put it, "the move from Lincoln Laboratory was not so much a radical change as a shift to a different perspective from which to view management" (Forrester 1991, 13).

When the Sloan School of Management opened in 1952, it built its curriculum from an undergraduate management course that had existed since 1914. It was during 1914 that David R. Dewey, brother of philosopher John Dewey and professor of economics at MIT, developed a course in engineering administration (Killian 1985). This course served as the basis for exploring Alfred Sloan's interest in developing a management school within a technical institution, rather than the usual formula of connecting it to a liberal arts institution, as was the case with Harvard University or the University of Chicago.

During the first four years of existence, the new management school was built along traditional lines, including courses in marketing, production, and accounting, without any particularly distinctive curriculum. As Forrester commented in an interview, "they had not yet done anything about the vision that had led Alfred Sloan to give them the money for a management school." Forrester's work, as he understood it, was to explore the ties between management and the more technical side of MIT. He proposed to do so by either "pushing forward the field of Operations Research or the use of digital computers for the handling of management information." He devoted the first year of his new job, with the dean's blessing, to exploring why he was there. After some time he concluded that pushing both the field of OR and computerized management of information was the wrong thing to do. Forrester felt that, although OR was "worthwhile," it was not dealing with the broad theme of the "reasons for the success and failure of corporations."[14] As far as management information, Forrester felt that he had done all that he could at that time. This is no small claim, since through his contribution to SAGE he was involved in a project for a network of computers that was to undergird an antimissile defense system for the whole of North America. Furthermore, he felt that the big corporations such as banks, insurance companies, and manufacturing were well launched in the art of using computer systems to manage their operations and that the possibility of making cutting-edge contributions was not very great.

His idea for a new approach came later that first year as a result of the "process I have always been engaged in, that is, to couple what I was doing to the real world."[15] The "real world" presented itself to Forrester when he had some conversations with managers from General Electric. As he recounts:

They were puzzled by why their household appliance plants in Kentucky
sometimes worked at full capacity with overtime and then two or three
years later, half the people would be laid off. It was easy enough to say
that business cycles caused fluctuating demand, but that explanation was
not entirely convincing. (Forrester 1991, 14)

Forrester decided to look into the operation to ascertain how the
managers were able to make decisions. He found out that they made
decisions on the basis of such things as incoming orders, order backlogs
to be filled, size of inventory, number of employees, and so forth. After
getting a sense of the functioning of the factory, Forrester started keeping
columns on a notebook page with information about employees, production rates, and inventories, and for any one line in a given week he started
asking the question, "Given this situation, what will people do? Hire more
people, produce more, layoff people? On the basis of what they do, you
would have the conditions for the following week."[16]
After doing some calculations in his notebook, Forrester came to
the conclusion that what they had at the factory was a clear case of a
feedback system: "Production affects inventory, inventory affects employment, employment affects production, and you find that this system is
unstable even if you have constant consumer demand."[17] His calculations
revealed an "internal structure and policies . . . that tended toward unstable behavior. Even with constant incoming orders, employment instability
could result from commonly used decision-making policies" (Forrester
1991, 14). This study of the Kentucky factories was published by Forrester
as an article in the *Harvard Business Review* of July–August 1958.
Intrinsic to Forrester's ideas is the process of seeing the company
as a dynamic system. In fact, when analyzed through System Dynamics,
the company is disassembled and reassembled. The company needs to
be recognized, according to Forrester,

> not as a collection of separate functions but as a system in which the
> *flows* of information, materials, manpower, capital equipment, and
> money set up forces that determine the basic tendencies toward growth,
> fluctuation, and decline. I want to emphasize the idea of movement
> here because it is not just the simple three-dimensional relationships of
> functions that counts, but the constant ebb and flow of change in these
> functions—their relationships as *dynamic* activities. (Forrester 1958, 52)

The ultimate goal of System Dynamics, still called "Industrial Dynamics" at the time, was the production of a model of the system, such
as a corporation, that would allow decision makers to have a tool for

institutional self-reflection in real time. Forrester's stress on the constant flow of change distinguishes his modeling from the more static and narrow traditional models of management that saw the corporation as a fixed aggregation of functions. In Forrester's system the corporation becomes more fluid and elastic. It loses its character as simply a static space within which bureaucratic functions are achieved. The corporation as fixed space within which personnel works to achieve specific functions gives way to the corporation as a stream of personnel, capital, materials, and equipment that needs to be managed. This management needs to be dynamic and in real time. The emphasis on real time, which Forrester developed while involved in the Whirlwind and SAGE projects, becomes central to System Dynamics in its modeling and simulating features.

The way Forrester has come to look at systems was largely shaped by his experiences during World War II and its aftermath. This particular way of thinking, first honed when he worked as an engineer, later became embodied in his management ideas and in the system of equations that serve as the backbone for computer simulation in System Dynamics. Forrester himself has something to say about this in his book *Industrial Dynamics* (1961), comprised largely of the Kentucky factory study and two other cases. In the first chapter of *Industrial Dynamics,* he discusses the four foundations of his systems thinking. One of them, and possibly the most important in terms of his work, is the development of the theory and practice of information-feedback systems, or servomechanisms. Forrester provides a broad definition of "servomechanisms." He writes: "An information-feedback system exists whenever the environment leads to a decision that results in action which affects the environment and thereby influences future decisions." This definition, according to Forrester, applies not only to "conscious and subconscious" decisions made by humans, but to "mechanical decisions made by devices." He provides three examples of information-feedback systems: a thermostat turning a furnace on or off according to the information it receives, a person correcting her balance if sensing she is about to fall, and a profitable industry attracting competitors until profit margin is reduced and competitors stop entering the field. According to Forrester, "all of these are information-feedback control loops. The regenerative process is continuous, and new results lead to new decisions which keep the system in continuous motion" (Forrester 1961, 14–15).

Information-feedback systems not only have the quality of always being dynamic and continuous, they also share three other characteristics, namely, structure, delays, and amplification. In Forrester's words:

The structure of a system tells how the parts are related to one another. Delays always exist in the availability of information, in making decisions based on the information, and in taking action on the decisions. Amplification usually exists throughout such systems, especially in the decision policies of our industrial and social systems. Amplification is manifested when an action is more forceful than might at first seem to be implied by the information inputs in governing decisions. (Forrester 1961, 15)

The structure, delays, and amplifications of a system can be best captured with the methodology of System Dynamics by modeling the system in question. This modeling depends on the gathering of information about the system from the participants in that system through their verbal descriptions. At least a portion of System Dynamics involves a sort of mini-ethnographic research project where information is gathered. Eventually, a model is made tracing the information-feedback loops that connect specific decisions to actions which, in turn, generate information that generates new decisions. This model is in turn reviewed by participants to make sure that it is seen as an adequate representation of the actual system. This modeling is done with the use of equations that capture the verbal description (Forrester 1961).

This modeling, which constitutes the second foundation of System Dynamics, has several purposes. The most important is to get at the endogenous structure of the system under study to see how the different elements of the system really relate to one another, and to experiment with changing relations within the system when different decisions are included. This modeling and simulating approach is far superior, in Forrester's view, to purely quantitative modeling, which cannot capture the complexity of large business operations.

A third foundation of System Dynamics, as seen by Forrester, is derived from the "automating of military tactical operations." He explains:

During WWII, fire-control prediction decisions were made automatically by machine, but before 1950 there was almost no acceptance of automatic threat evaluation, weapon selection, friend or foe identification, alerting of forces, or target assignment. In a mere ten years these automatic decisions were pioneered, accepted, and put into practice. In so doing, it was necessary to interpret the "tactical judgment and experience" of military decision making into formal rules and procedures. (Forrester 1961, 17)

This third foundation, of interpreting and reducing human judgments in such a way that they can be formalized and operationalized in a simulation process, has come to be the foundation not only of System Dynamics but of much of the new thinking in the field of management in the last forty years.

The fourth foundation of System Dynamics, in Forrester's view, is the digital computer. These machines facilitate the simulation of complex systems because of their ability to handle many equations. Furthermore, their increasing popularity makes their cost small in comparison to other operational costs. This combination of comparatively low cost and ability to deal with complex systems makes computers indispensable to "the rate of progress in understanding system dynamics" (Forrester 1961, 19).

The four foundations that compose System Dynamics emerged largely from the crucible of World War II. The concept of information-feedback loops came, at least in Forrester's experience, out of his work with servomechanisms. His awareness of automation of military decision making was at the heart of the SAGE system. The impetus of that program—to develop computers that could operate in real time—was necessitated by the need to automate military decision making during a potential air attack on the United States. The emphasis on simulation was derived from Forrester's goal of capturing the dynamic nature of systems. His experience with servomechanisms in general, and the flight simulator in particular, led him to feel that in order to capture the structure of a system one has to also capture its dynamic nature. His experience with the flight simulator gave him confidence that this could be done, provided adequate equations and computer power were available. The fourth foundation of System Dynamics, namely, its reliance on computers, emerged largely out of Forrester's long experience with Whirlwind and SAGE when he experienced firsthand the power to be had with the use of these machines.

The nature of Forrester's contribution can be seen in his ability to integrate many of these disparate experiences, concepts, and machines into a coherent set of tools for management. The emergence of this set of tools, which Forrester ultimately called System Dynamics, was historically contingent on his experiences and the exigencies of the moment.

From Industrial to Urban Dynamics to World Dynamics

While there are nuanced differences in the way that Forrester integrated System Dynamics with the subjects under study (such as corporations, cities, national economies, or the whole planet), the underlying theme is one of methodological and normative continuity. This continuity is provided by his conviction that all technical, social, and biological activities share similar systemic structures. This fact makes all areas, according to Forrester, amenable to analysis by looking at their systemic mechanisms and overall systemic structure. This notion, of course, he shared with von Neumann, Wiener, the Odum brothers, and others operating under the sign of cybernetics and systems analysis. In one of our interviews, Forrester commented that the idea of seeing all systems, whether technical, social or ecological, as sharing similar structure came to him over a twenty-year span, between the 1940s and 1960s, with his move to the Sloan School as a particularly important time in that process.[18]

Forrester's model in *World Dynamics* mapped out the interrelationships between six main variables that he saw as fundamentally constituting the endogenous structure of the world system. These variables are natural resources, population, geographic space, food production, pollution, and capital investment. Throughout the book, Forrester tended to conflate these six variables into four: natural resources, capital investment, population, and pollution. Values for these variables were then plugged into the model, though the values that Forrester used had very little empirical confirmation. In addition, the structure of the model itself has two fundamental categories: levels and rates. The levels are "the accumulations (integrations) within the system. The rates are the flows that cause the levels to change" (Forrester 1973b, 18).

An essential component of the system is the feedback loop. As Forrester defines it, this loop is the "closed path that connects an action to its effect on the surrounding condition, and these resulting conditions in turn come back as information to influence further action" (Forrester 1973b, 17). These feedback loops can be positive or negative. An example of a positive one is the "vicious circle" of exponential growth. A thermostat that tends toward homeostasis is an example of negative feedback.

Forrester states that activity in all systems is characterized by feedback loops. This characteristic lends systems a "counterintuitive" quality for which the human mind is not properly equipped. As he argues in a paper on the behavior of social systems:

> The human mind is not adapted to interpreting how social systems work. Our social systems belong to the class called multiple-loop nonlinear feedback systems. In the long history of human evolution it has not been necessary for man to understand these systems until very recent historical times. Evolutionary processes have not given us the mental skill needed to interpret the dynamic behavior of the systems of which we have now become a part. (Forrester 1973b, 5)

What allows humans to overcome this evolutionary limitation, according to Forrester, is the computer, or more precisely, computer modeling. He points out that nations would not attempt complex technological feats, such as sending a spacecraft to the moon, without first running computer simulations of the possible space trajectories and general functioning of the equipment. Likewise, he asks, "Why do we not use the same approach of making models of social systems and conducting laboratory experiments on those models before we try new laws and government programs in real life?" (Forrester 1973b, 126).

The World Dynamics model that Forrester developed combines the notions of "counterintuitive" feedback loops with a global perspective. It is a model that allows human beings to see natural laws and forces at work on the planetary level. He states:

> Within one lifetime, dormant forces within the world system can exert themselves and take control. Falling food supply, rising pollution, and decreasing space per person are on the verge of combining to generate pressures great enough to reduce birth rate and increase death rates. When ultimate limits are approached, negative forces in the system gather strength until they stop the growth processes that had previously been in control. In one brief moment of time the world finds that the apparent law of exponential growth fails as the complete description of nature. Other fundamental laws of nature and the social system have been lying in wait until their time has come. Forces within the world system must and will rise far enough to suppress the power of growth. (Forrester 1973b, 5)

The most dramatic and essential conclusion that Forrester drew out of running the model was that unless radical efforts were made to achieve a sustainable level of industrial and population growth, the world would experience a depletion of natural resources and, after rapid exponential growth, a sharp decline of population.

In talking about *World Dynamics*, I am getting ahead of the story. The next chapter discusses the historical events that led Forrester to conceive of modeling the whole planet. While the process involved adapting and extending the methodologies used to model corporations and cities, it was not a given that this was going to be the next use of System Dynamics. In fact, the way in which Forrester came to conceive of *World Dynamics* and the conclusions that the book embodied were the result of a complex process of serendipity and historical contingency. I now turn to that story.

Notes

1. Jay Forrester, interview by the author, December 12, 1991.

2. For example, Gordon Brown recently contributed financial support to a project dear to Forrester's heart: teaching System Dynamics to high school students. The pilot project has been established in New Mexico, where Brown is spending his retirement years. Brown was the person who introduced Susan Swett to Forrester. She and Forrester eventually married in July of 1946. Brown wrote the introduction to Forrester's *Collected Papers* (1975), and Forrester dedicated his *World Dynamics* (1973b, 2d ed.) to Brown, his lifelong mentor.

3. Forrester, interview by the author, May 20, 1994.

4. De Florez, an MIT graduate, was another one of those amazing figures who shaped technoscientific practices both during World War II and in the postwar period. At the time of ASCA, de Florez was director of the Special Devices Division of the Bureau of Aeronautics for the navy. Before joining the navy in 1939 he had developed an international reputation as an expert in aviation (Redmond and Smith 1980). During his tenure at the Special Devices Division he championed the development of simulation devices for training and designing. After the war, Admiral de Florez became the first director of research for the Central Intelligence Agency (Sapolsky 1990). He was remembered around MIT for visiting in his seaplane. His was the only permit issued to "land" on the Charles River next to MIT (Sapolsky 1990; Forrester 1991).

5. While this particular project failed, the development of flight simulators continued apace. By the 1960s with the addition of image generation systems, used to simulate features such as airports, flight simulators were fully developed. It was during World War II, though, that most components of the flight simulator were put together. In fact, World War II remains "the watershed of simulator development" (Rolfe and Staples 1986, 35).

6. The Whirlwind project itself went through several transformations until in its last stage it was seen as a "laboratory for exploring how digital computers could serve as military combat information centers" (Forrester 1991, 11). The history of this project is not the focus of my work, but an excellent history can be found in Redmond and Smith (1980). See also Edwards (1996).

7. In chapter 2 I mentioned Haraway's article exploring E. O. Wilson's early research and its dependence on system conceptualizations. It is worth highlighting here the way in which the design of a flight simulator encouraged

similar conceptualizations, in this case the human pilot and the simulator being thought of as a smoothly integrated system. It is in the aggregation of multiple examples such as those of Wilson and the flight simulator that the boundaries between the organic and the technical were largely set aside, giving rise to the systems language and practices that facilitate the treatment of all things within the same field of analysis.

8. Everett continued working in Whirlwind-related projects after Forrester left the computing field in 1956. Everett would eventually become the president of the MITRE Corporation, an offshoot of Servomechanisms Laboratory and Whirlwind.

9. Gordon Brown as quoted in Wildes and Lindgren (1985, 218).

10. The history of numerically controlled machines, including the MIT-Parsons relation, is discussed by Noble in his *Forces of Production* (1984). See also *Numerical Control: Making a New Technology* by J. Francis Reintjes (1991).

11. A discussion of many aspects of the history of SAGE can be found in Robert R. Everett, ed., special issue: "Sage," *Annals of the History of Computing* 5, no. 4 (October 1983).

12. The original grant from the Sloan Foundation was for $5,250,000.

13. Forrester, interview by the author, December 12, 1991.

14. Ibid.

15. Ibid.

16. Ibid.

17. Ibid.

18. Forrester, interview by the author, May 20, 1994.

Four

The Club of Rome

Cybernetics returns. For the first time in history, the human world is face to face with the worldwide world.

—*Michel Serres*[1]

The emergence of a new form of governmentality based on discourses and practices concerned with the planet as a whole is not the only element that constitutes the new order of things. In addition, we see the emergence of new institutions, many of which actually produce and carry out the practices associated with managing the whole planet. A quintessential exemplar of such an institution is the Club of Rome. Its very purpose is to provide the framework for the production of knowledge that transcends national borders and focuses on the intertwining of various problems. An early articulation of the purpose of the Club of Rome stated that it was intended to

> a. Contribute toward an understanding of the problems of modern society considered as an ensemble, and to the analysis of the dynamics, interdependencies, interactions, and overlapping that characterize this ensemble, concentrating particularly on those aspects that concern all or large sections of mankind.
> b. To heighten the awareness that this complex of tangled, changing, and difficult problems constitutes, over and above all political, racial or economic frontiers, an unprecedented threat to all peoples, and must therefore be attacked by the multinational and transnational mobilizing of human and material resources.[2]

The objectives mentioned above clearly suggest how intimately related the very existence of the Club of Rome and its goals are to conceptions of spaces of governmentality larger and possibly different from nation-states. The emergence of global spaces and the Club of Rome, as well as the emergence of similar organizations, is mutually constitutive. The Club of Rome is, of course, not the only organization that articulates its purposes in relation to global spaces. Research organizations like World-watch Institute (note the global perspective reflected in the very name) or the Center for the Study of Global Governance at the London School of Economics are just two examples of the many organizations, institutes, and think tanks that inhabit those spaces through their research, advocacy practices, and discursive strategies.

In many ways the emergence of these organizations can be traced, at least indirectly, to novel institutional arrangements extended, made possible, or invented during and immediately after World War II. It is during this period that we see a geometric growth of private or quasi-public institutions that focus in the crafting of public policy through advice and/or advocacy vis-à-vis governmental institutions. This era pro-pelled the growth of the think tank. The very name is associated with World War II military jargon. It referred, according to historian James A. Smith, to "a secure room where military plans and strategies could be discussed" (James A. Smith 1991, xiii).

The RAND Corporation, which was established toward the end of the war as an offshoot of the then Douglas Aircraft Corporation, is the premier exemplar of these new institutions. RAND was established originally by Douglas Aircraft at the request of the air force, which saw the need to retain the expertise of civilian scientists and engineers after the immediate demands of winning World War II were met. The Air Force High Command had been particularly impressed with the contributions that the use of OR had made to air force activities. After two years of existing as a project within Douglas Aircraft, RAND became an independent nonprofit research organization in 1947 and remains a thriving entity headquartered in Santa Monica, California.[3]

After the war, the American federal government became a prodi-gious consumer of the work of outside experts and has remained so to this day. The number of organizations feeding this voracious appetite has grown exponentially since the war and has branched out to cover the needs of other governments, international institutions, and private corporations. The importance of these institutions lies not so much with their knowledge production, though it is quite vast and important, but with their often tacit function of building bridges between different branches of government, universities, international organizations, and the private sector. It is this thick web of interrelations that is at the core of the postwar era, particularly in the Western world.

The Club of Rome, with its membership composed of industrialists and highly placed public bureaucrats, can be seen as yet one more of these organizations, providing an ongoing and emergent structure of governmentality and knowledge production.

The Club of Rome and the Emergence of Global Spaces

> It all started when, by accident, a Soviet scientist colleague, flipping through a magazine in an airport waiting lounge, came across a report of a speech made by Aurelio Peccei at a conference of industrialists in Buenos Aires. Impressed by what he saw, he had a copy sent to me at OECD with the terse comment "this is what you should be thinking about." At the time, I had never heard of Peccei. I made inquiries about him and wrote him immediately asking for a meeting. A week or so later we had our first conversation.[4]

With these words, Alexander King, then director general for scientific affairs at the Organization for Economic Cooperation and Development (OECD), narrates how in 1967 he came to meet Aurelio Peccei. This meeting is credited by both King and Peccei as the first step in founding what would come to be known as the Club of Rome.

It is somewhat ironic that the beginnings of an elite association of Western industrialists and technocrats came about in the accidental flipping through a magazine by a scientist of the then Soviet Union. The scientist in question was Jermen M. Gvishiani, the son-in-law of Aleksey Nikolayevich Kosygin, the foreign minister of the Soviet Union at the time.[5] As a story of origins, it illustrates the historically contingent nature of institutions and practices, however solid and inevitable they may appear later on.

The talk that caught the attention of Gvishiani and later King was originally given by Peccei at the National Military College in Buenos Aires in September of 1965. The talk, entitled "The Challenge of the 1970s for the World of Today," was sponsored by Atlantic Development of Latin America (ADELA), a privately owned institution created by several multinational corporations from Europe and the United States to channel investments into the Latin American private sector. In that speech Peccei, a maverick Italian industrialist then affiliated with the FIAT car company, outlined some of his concerns about the growing gap in wealth between the industrialized nations of the North and those of the South.[6] He expressed particular concern about the impact of what he called the "second industrial revolution." This "revolution" consisted in the systematic application of information technologies, especially computers, to industrial and managerial processes. For Peccei, the second industrial revolution, led by the United States, was quite problematic. His concern was that this new form of production would not only deepen but also accelerate the divisions between the rich and the poor (Peccei 1965). He advocated accelerated transfer of technology and managerial know-how to the "underdeveloped" regions. He felt that Latin America could turn into a model of development to be eventually exported to Africa and the poorer regions of Asia.

Peccei's talk reflected the broad concerns that he was developing at the time. They are remarkable when compared to the normal concerns of corporate managers. His concern about the poverty gap, for example, was not something that exercised many corporate managers then, or now, for that matter. The talk displays the broad reading and learning that characterizes Peccei's thought. In it he refers to a wide variety of studies and reports. For example, he quotes a research report from the Stanford Research Institute (now known as SRI International) on the probable state of the world in 1975. He also makes reference to reports by the UN Conference on Trade and Development, talks by the chairman of the International Industrial Conference, research projects by MIT, and figures published in the *Bulletin of the European Economic Community* and other such publications. The talk reflects an omnivorous mind and a desire to keep informed and to keep learning.

Soon after King and Peccei first met, they decided to put together a group of approximately forty friends and colleagues to discuss forming a "non-organization" to reflect on world problems (Pauli 1987, 72). The idea of a "non-organization" expressed their desire to build a structure to facilitate discussions without the encumbrance of a plodding bureaucracy and without having to spend an inordinate amount of time on political maneuvering. Prior to the meeting Peccei and King decided that a background paper on world problems would be useful (Peccei 1977). Following a suggestion by King, Peccei contacted Erich Jantsch, an Austrian astronomer turned consultant, and asked him to write such a paper (Peccei 1969a). Jantsch was then a consultant to the OECD and had recently published a book on technological forecasting under the auspices of that organization (Jantsch 1967).

Jantsch produced for the meeting organized by Peccei and King a paper entitled "A Tentative Framework for Initiating System-Wide Planning of World Scope" (Peccei 1969a). Whether this paper ended up being a help or a hindrance to the meeting is hard to say. In 1969, Peccei stated that "Dr. Jantsch did a splendid job" (Peccei 1969a, 251). A few years later his memories of that work were less charitable. He wrote that the background document was not "easily readable" (Peccei 1977, 63).

The First Meeting at the Accademia Nazionali dei Lincei

The first meeting of the forty or so luminaries from the technocratic and business elite of Europe occurred in Rome on April 6–7, 1968. Peccei had obtained financial support for this meeting from the Giovanni Agnelli Foundation, the charitable arm of the Agnelli family. The Agnellis are the founders and still largest stockholders of the FIAT corporation for which Peccei had worked many years.[7] The meeting took place at the Villa Farnesiano of the Accademia Nazionali dei Lincei. The Accademia, founded in 1603, was originally a gathering place for supporters of Galileo's ideas. The Villa Farnesina offered a beautiful and serene meeting place, decorated with paintings by Raphael, Sebastiano del Piombo, Giulio Romano, and other Renaissance painters (Peccei 1977). Apparently, neither the beauty of the place nor the Frascati wines that punctuated their meals were sufficient to make the meeting a success. "A partial success," observed Peccei of the meeting (Peccei 1969a, 251); "a monumental flop," adjudged King.[8] A meeting of the minds on how best to pursue an exploration of world problems did not occur. Members of the group spent many hours discussing such issues as the difference in meaning between the English word "system" and the French word "système." Peccei attributed part of the problem to the heterogeneous background of the participants and their commitment to the languages of their respective areas of expertise. He noted:

> As one can easily expect with such a new topic as world-scope planning, some theological and semantic battles could not be avoided, as well as a conceptual debate on the applicability of systems analysis and computer-based techniques to some of these elusive subject matters. (Peccei 1969a, 252)

In fact, the emergence of clear conceptualizations about the goals of the Club of Rome and possible methodologies to achieve those goals were always fraught with conflict, resistance, and accommodation. The process was far from painless. Ideas that were to become key to the Club of Rome projects were not automatically accepted, and the loftier aspirations of the club did not provide access to eternal truths. The "truths" of the global discourses, their connections to practices of cybernetics and computer simulation, emerged out of hard work, out of many efforts to hammer out a concrete statement or practice. Certainly, Peccei's description of that first meeting is a clear example of the process.

Even though the meeting ultimately failed in its larger goal of launching a new way of looking at the world, a continuing steering committee was formed, composed of Peccei, Alexander King, Erich Jantsch, Max Kohnstamm (a Dutch expert on international issues), Jean Saint-Geours (an economist and financial consultant), and Hugo Thiemann (then head of the Battelle Institute in Geneva).[9] This group met during a final dinner at Peccei's house and agreed to set up the Club of Rome, the name referring to the city where they first decided to become a "non-organization" (Pauli 1987, 72). It was this evening meeting that is credited as giving birth to the Club of Rome. The immediate goal of this "non-organization" was to maintain and develop further intra-European contacts in government and private industry and to explore further steps of action.

The OECD Bellagio Conference on Forecasting

Another important meeting affecting the ultimate direction of the Club of Rome occurred from October 27, 1968, to November 2, 1968. That meeting, at the Villa Sebelloni in Bellagio, Italy, was organized by the OECD to further discussion on forecasting and long-range planning by a selected group of experts.[10] Jantsch was a pivotal figure in organizing the conference, and it was probably his opinion and that of King in his capacity as director general for scientific affairs at OECD that carried the most weight in determining the list of invitees. It was at this conference that Forrester and Peccei met for the first time, although nothing in particular came from that meeting.[11] Forrester was eventually invited to become a member of the Club of Rome. That invitation, however, did not come until March of 1970.[12]

At the OECD Bellagio Conference, one of the speakers was Hasan Ozbekhan, then director of planning at the System Development Corporation (SDC). The SDC had originated from a research program which was exploring how man-machine systems affected collective organizational behavior. This program was begun in 1950 at the RAND Corporation in Santa Monica, California (Baum 1981). It was to be the kernel for the training program of computer programmers for the SAGE defense system. The training program soon became a monumental task, and it was decided to spin out both the original program and the SAGE training as a separate corporation in 1956.[13]

Ozbekhan gave a wide-ranging, 108-page presentation entitled "Toward a General Theory of Planning." In this presentation Ozbekhan listed twenty-eight "Continuous Critical Problems" (Ozbekhan 1969, 85). These included generalized poverty within affluence, hunger and malnutrition, environmental pollution, uncontrolled population growth, polarized military power, discrimination against minorities, and inadequate conception of world order. This list gives a flavor for the range of problems that Ozbekhan thought needed to be dealt with in terms of overarching systems. His presentation had a strong normative content and stressed the need for long-range planning that included a dynamic component, to accommodate the ever-changing human values. He argued that computer modeling would be a very good tool to conceptualize relations and embody the dynamic aspects of the system under study.

Forrester was also an active participant at the conference. He gave two talks. One of them focused on the need to rethink the corporation (Forrester 1969a). It was a rather visionary talk in that it discussed the need to decentralize decision making and to restructure management by making use of computer-driven information networks within the enterprise.[14] Forrester's main talk was prepared specifically for the Bellagio conference, though it included material that he had presented before. This talk focused on his specific ideas of Industrial Dynamics as applied to the city (Forrester 1969b). The presentation had a theoretical component (Industrial Dynamics), a specific methodology (computer modeling), and an example of how the system dynamics worked (modeling an urban area), together with some specific results from running the simulation. Those results were presented in the form of charts depicting the different trajectories and interactions of such variables as labor, worker housing, tax burdens, and unemployment. They tended to show that "many policies, intuitively appealing, politically attractive in the short run, and apparently humanitarian, may lead in the wrong direction" (Forrester 1969b, 250).

Peccei did not give a formal presentation at the Bellagio conference. As he himself acknowledged, he was not an expert at long-range planning and modeling. He was there at King's invitation more as an observer than as an active participant. Peccei did contribute, along with Forrester and others, a series of reflections on the conference. He argued that all participants and readers of the proceedings should agree that "we are going through a time of planetary emergency" due to the severity, complexity, and depth of the problems facing humanity. In this light, he went on to argue, "a radical change in our course and methods of managing the world (versus today's unmanagement-mismanagement) is clearly a very high priority" (Peccei 1969b, 518). Peccei's talk of "planetary emergency" and "managing the world" reflects the discourse of globality. It is a discourse that had not yet found a tool for practicing that vision even though it is evident at this point that Peccei saw some sort of long-range modeling as an "appropriate way to go." He had the feeling that the new vision of a worldwide "problematique"—the interconnectedness of world problems—demanded new tools, new ways of observing, planning, and managing. In other words, global problems required new tools of practice to become visible and manageable. Peccei's somewhat pithy comments also reflected his move away from a discourse focused solely on "man" and the problems of human society per se toward a more inclusive vision in which humanity is seen as part of a larger system.

At the end of 1968, Peccei and the other members of the steering committee of the Club of Rome were confronted with at least two large problems: how to "jump-start" a worldwide dialogue about global problems by "enrolling," in Bruno Latour's sense of the term, the media as well as elite political and business groups into paying close attention to the need for new visions and understandings of world problems (Latour 1987). In other words, they had to convince enough people in positions of authority of the need for a new perspective, a new mode of perception. The second large and related problem was to come up with a methodology that would make the interrelatedness of world problems transparent to all.

Before exploring how the Club of Rome eventually solved these problems, let us look more closely at Peccei, the main dynamo behind the ultimate success of the club. His personality, character, and ideas animated much of what the organization did until his untimely passing in March of 1984. Therefore, a discussion of his work and ideas can help to illuminate the ways in which the transition from a worldview based on the notions of "man," in the Enlightenment sense, to a worldview more oriented toward the planet itself emerges within an individual. Peccei's ideas reveal the ambiguities and complexities that are opened when a global perspective is accepted. Looking at Peccei also helps to explain the methodological choices that the Club of Rome made as they sought to meet their goals.

Aurelio Peccei

Peccei was interested in planning from very early in his life. His economics dissertation at the University of Turin was on Vladimir Lenin's first five-year plan (Pauli 1987).[15] He explored this particular theme after being attracted by the energy of the communist revolution in the Soviet Union. In fact he took it upon himself to learn Russian and became "fairly fluent" in that language. At that point in his life, he admired Karl Marx's thinking and had since valued it as a legacy that, together with that of others, should be reinterpreted "to suit the new conditions of our time" (Peccei 1977, 3). While this interest in socialist economics did not always endear him to his colleagues at FIAT, it no doubt facilitated his role in mediating between East and West during the long years of the cold war. Peccei played well the role of mediator as illustrated in the setting up of the IIASA, discussed further below.

It was not until his job as director of FIAT's efforts to establish subsidiaries in Latin America that Peccei exercised his lifelong interest in planning.[16] As part of that job, he drafted a comprehensive five-year plan, the first of its kind at any of the divisions of FIAT (Pauli 1987). It is evident that Peccei's thinking was very much imbued with forecasting and rationalizing strategies, as well as with a humanism which tended to perceive quantifying and anticipatory mental strategies as tools rather than as an exhaustive way of apprehending the world.[17] It was this thinking that he wanted to see embodied in the work of the Club of Rome.

Although he would eventually become identified with a global outlook, Peccei was not born with such a perspective. It emerged as part of his outlook over the course of many years. The organizational focus of his thinking remained within the confines of the corporation and the nation-state for most of his life—until the early to mid-1960s. In 1964 Peccei began to show increasing concern about the ultimate goal of a developmentist and industrializing strategy for what were then known as "underdeveloped" societies.[18] Dr. Ricardo Diez-Hochtleitner, current president of the Club of Rome, remembers that when he first met Peccei in the early 1960s Peccei was already raising questions about the gaps between the rich and the poor and between the United States and Europe in the areas of technological progress. He also recalls Peccei's asking questions about the organization of education around the world.[19] These concerns were enlarged by the rapid growth that the United States was experiencing vis-à-vis Europe. Peccei's talk in Buenos Aires, mentioned above, reflected this new concern. Eventually he would also become very concerned about China and its potential to become a dominating economic and military power (Peccei 1969a). These concerns were not exclusive to Peccei; they were shared by many members of the European technocratic elites. A clear embodiment of that concern was the publication, in 1968, of Jean-Jacques Servan-Schreiber's *The American Challenge*. This book, which was a bestseller in several countries in Europe, expressed the fear that U.S. wealth production, anchored in information technologies, would leave Europe behind. Still, it can be said that Peccei was an early expositor of these concerns.

Of course, Peccei's thinking underwent many transformations even after he started to become interested in looking at the wider problems of humanity. Sometimes he is depicted as unproblematically being ahead of the curve, waiting patiently until at least a few more people catch up with him.[20] Certainly, he was broadly foresighted in his thinking and possessed a charismatic personality. Still, Peccei can be distinguished more for his willingness to learn than from his foresight per se. His global vision emerged in time through his interactions with many other people and his encounter with the specific tools of computer modeling. He was very attracted, for example, by the work done at the RAND Corporation and SDC. While Peccei's outlook was largely humanistic, he was drawn to the potential of systems thinking and practices as a way of conceptualizing and explaining the interrelatedness of disparate problems.

The shift in Peccei's outlook, from thinking about international problems to thinking about global problems, was the result of many years of experience in the developing world and a growing sense that problems were not being addressed in their manifold complexity. It was probably not until 1968 that Peccei began to articulate elements of what I call a discourse of globality in a systematic manner. These elements include a focusing on problems that transcend political frontiers, that see human activity as integral to the biosphere, that announce a general concern for the fitness of the planet as a whole, that require a discussion of the fate of humanity as an undifferentiated whole, and that focus attention on the ultimate finitude of Earth rather than, for example, the development of national economies.

In Peccei we can see articulations of this perceptual shift in his book *The Chasm Ahead* (1969a). In this book, originally published in 1969, he calls for a "new approach." This new approach should be guided, according to him, by two concepts:

> One may be built around the concept of the necessity of understanding far better the changed relations between man, society, and environment in this period of revolutionary transformation. The other is descending from the interference of all questions one with the other, and hence the necessity that each problem or family of problems be analyzed in a wider context—which for the larger problems is a world context. All of them point out the world's oneness, its being about our condition and the future. (Peccei 1969a, 137)

In other words, Peccei claims the necessity of a global context if the ubiquitous changes in the relations between human society and the environment are to be understood.

Earlier in the same book, he quotes with approbation from the paper that Jantsch had prepared for the exploratory meeting that Peccei had organized at the Accademia dei Lincei. Jantsch, speaking of the "growth complex" of Western societies, stated:

> The absolute level attained by this uncontrolled growth implies a high inertia of the dynamic system and reduced flexibility for change. It also dawns on us now that there is no inherent cybernetics in the system, no self-regulating automatism of macro processes; the cybernetic element in the evolution of our planet is man himself and his capacity for actively shaping the future. (Peccei 1969a, 136)

Amplifying on that quotation, Peccei admonishes:

> Now that he has created forces and cycles which compete and interfere with those of nature itself, but which possess no in-built regulating mechanism, man—to avoid economic, social, political, demographic or ecological debacle, and guide his destiny—has himself to be the cybernetic or regulating element of all man-influenced processes. (Peccei 1969a, 136)

It is worth highlighting in these quotations the planetary-level language and the connection of that language with cybernetic thinking. Again, it is cybernetic and systems thinking that facilitates a planetary conceptualization.

In an article that Peccei wrote in February of 1969, he included a series of illustrations of his ideas (Peccei 1969a). One of those illustrations gives us a vivid pictorial representation of what I have called a discourse of globality. It presents technology at the center of a system that includes man, society, and nature, all enveloped within the biosphere. It suggests, with its roundness, the whole earth. The terrain of perception is the earth as a whole, composed of a system of interrelationships that human beings helped to create and now need to regulate. Nature itself is depicted as a crucial element of the system, with the biosphere serving as the overarching component of the system. The net effect is the representation of what can be called a new "field of intervention." By "field of intervention" I mean the conceptualization of a space made amenable to intervention and management. This intervention may occur through many processes, including, for example, scientific experiments, political accords, military power, and economic production. These interventions, however, cannot literally be achieved without defining the field within which they are to occur.

While Peccei and members of the steering committee of the Club of Rome held high-level meetings and gave many talks and interviews, they felt that their message was not "getting through." It was, as Peccei put it, "as if the global problems we were ventilating concerned another planet . . . our words carried no more weight than the Pope's homilies, the UN Secretary General U Thant's admonitions, or the warnings of concerned scholars and thinkers" (Peccei 1977, 67).

The conclusions to which the Club of Rome members came, after many months of heightened activity, were stated succinctly by Peccei: "We clearly understood that, to rivet people's attention on issues apparently so remote from the immediate interests of their life, a radical change of methods and means of communication was necessary" (Peccei 1977, 67).

Notes

1. Michel Serres, "The Natural Contract" (1992, 15).

2. These objectives are listed in annex 1 of the original proposal of the Club of Rome to the Volkswagen Foundation to obtain financial support to start the study of problems on a worldwide basis. The proposal, entitled "The Predicament of Mankind: Quest for Structured Responses to Growing World-Wide Complexities and Uncertainties," was prepared by Hasan Ozbekhan. I found a copy of this proposal in the personal papers of Carroll Wilson at the MIT archives. His papers are identified as MC29. This particular document is located in MC29, box 55, folder 2113 (MC29-55-2113). I discuss this proposal and its significance below. Ozbekhan and Carroll Wilson are also further discussed below.

3. The early history of RAND is covered in Bruce L. R. Smith, *The RAND Corporation* (1966).

4. Alexander King, quoted in "Cool Catalyst," a profile of him (Brabyn 1972, 390). See also Peccei (1977). Peter Moll, in his excellent book on the impact of the Club of Rome on the development of future studies, suggests that the beginnings of the Club of Rome were even more tortuous. Apparently, Peccei's talk was transcribed and eventually printed in a UN publication. It was this publication that Jermen Gvishiani picked up. He became interested in the article and its author, Aurelio Peccei. King, quoted by Moll, takes the narrative from that point: "He (Gvishiani) sent a copy of the paper to his American colleague of the UN committee asking him to find out who this Peccei was and to put them in touch. The American was Carroll Wilson, a good friend of mine who later became a member of the Club. He had never heard of Peccei either, so sent the material on to me in Paris with the request to do my best to locate the Italian" (King, quoted in Moll [1991, 61]).

5. The information on Gvishiani's family connection is from Ricardo Diez-Hochleitner, the current president of the Club of Rome, interview by the author, April 1, 1993.

6. A copy of this speech can be found in the Aurelio Peccei Papers located in Rome. The papers are organized in five files, with each document within a particular file numbered. The copy of the Buenos Aires speech is located at the beginning of the first file and is identified as "pre-numero 2." It should be added that unfortunately most of the early papers about the Club of Rome (1968–76)

were disposed of by Peccei a few years ago when he had to quickly move out of an office he was given at Italconsult (an engineering and management consulting firm). A sudden change of leadership in that company in 1982 required that Peccei vacate the office within a week. In anger he left that same day, packing only the documents of immediate importance. Most historical records were not included. This fact has, of course, been a problem for those looking at historical aspects of the Club of Rome. Gunter Pauli, who wrote a 1987 biography of Peccei, and Peter Moll, who wrote a 1991 book about the development of future studies focusing on the Club of Rome, were faced with this same problem. I was very fortunate, however, in finding a great deal of archival material relating to the early years, particularly to the whole process culminating in the *Limits to Growth* project, in an unsuspected location: the archival papers of both Carroll Louis Wilson and Gordon Stanley Brown located in the MIT archives. Wilson was a very early member of the executive committee of the Club of Rome and even had communications with Peccei prior to his formal membership. Peccei forwarded to him copies of letters that he had sent to other people, as well as all internal reports about the ongoing activities of the Club of Rome. The Gordon Stanley Brown Papers have copies of virtually all of the correspondence that Forrester had with and about the Club of Rome, as well as other materials relating to the club. This happy circumstance can be explained by the close friendship between Forrester and Brown. Furthermore, Brown had a very keen interest in developments related to Forrester's Industrial Dynamics group at MIT. Thus, Forrester always kept Brown fully informed of developments concerning his career.

7. The Agnelli family owns about 36 percent of the total FIAT shares outstanding, according to an article in *The Economist,* January 29, 1994, p. 65.

8. King is quoted in Brabyn (1972), p. 391.

9. The Battelle Institute in Geneva is connected to the Battelle Institute in Columbus, Ohio.

10. A collection of the papers presented at that conference, as well as the list of participants and contributors, can be found in Jantsch (1969). The participants included King, Peccei, Ozbekhan, Dennis Gabor, Forrester, and Jantsch. At different times they would all be members of the Club of Rome. The Villa Sebelloni was made available for the purposes of the meeting by the Rockefeller Foundation. In my interview with him, Forrester attributed his being invited to the conference to the favorable impact of his recently published book, *Urban Dynamics* (Forrester 1969c).

11. Forrester, interview by the author, December 12, 1991.

12. A copy of Peccei's letter inviting Forrester to join can be found in the Gordon Stanley Brown Papers, MC24-17-683.

13. It is important to highlight again a point made in chapter 2—namely, that a thick web of discourse, practices, and relations emerged out of World War II around notions of OR and systems thinking. The fact that both Forrester and Ozbekhan were connected with the same project, coming to it through different avenues and from opposite geographic points, illustrates the crisscrossing of these techniques throughout the world, especially in the West. The tremendous

importance of SAGE in these developments should not be underestimated. It was the technocratic womb for many material and mental techniques that today constitute the world of computer technology and systems thinking.

14. This presentation was not new. Forrester had published essentially the same paper in *Sloan Management Review* (1965). The talk was nonetheless visionary in that it articulated what has now become the hot issue in management restructuring. A current discussion of the wide-ranging impact of information networks in the restructuring of corporations can be found in Taylor and Van Every (1993).

15. Contrary to the fashion of the times, when Turin was under a fascist administration, Peccei courageously defended his thesis in a white shirt rather than in the customary black. During the war Peccei was involved with the anti-fascist resistance. It was a stand that did not endear him to the powerful Agnelli family, who on the whole were very supportive of the Mussolini regime. Peccei came perilously close to losing his life for his resistance activities. At one point he was captured and was scheduled for execution. It was a fortuitous set of circumstances that prevented his execution (Peccei 1977).

16. Peccei had first joined FIAT in 1927. His early job involved selling FIAT's planes in China, a job cut short by the Chinese communist revolution.

17. The humanistic strain in Peccei's thinking is evident in many of his writings (see, e.g., his book *The Human Quality* [1977]), and it was confirmed to me by Roberto Peccei, one of Aurelio Peccei's two sons, in a phone conversation. Roberto Peccei is currently chairperson of the physics department at UCLA. His brother, Riccardo Peccei, teaches industrial relations at the London School of Economics.

18. The notions of "development" and "modernization" need to be seen as historically constructed conceptualizations. They establish descriptions and classifications that make certain types of intervention, such as international financing of "development" projects, appear natural and without any political content. In other words, "developmentism" is a historically specific way of organizing the world and not, as the literature from within the development field itself would have it, a mere objective description of the world. See Escobar (1987) and Ferguson (1990). I make a similar point for the global spaces I am talking about in this book.

19. Diez-Hochleitner, at that time, was involved as a consultant to international bodies and to some national governments on matters of educational policy.

20. His biographer Pauli (1987) sometimes falls into this pattern.

Five

The World Problematique Encounters System Dynamics

At a meeting of the Club of Rome in September of 1969, several members of the organization—which now included Eduard Pestel, President of the Institute for Systems Analysis and Prognosis at the University of Hanover; Conrad Waddington, a biologist; and Detlev Bronk, President Emeritus of the U.S. Academy of Science—reached a consensus on the next steps. Peccei stated that

> the consensus was that the most promising way to attain our objectives was that of presenting and analyzing the world problematique by the systematic use of global models. Never before had mathematical models been used to describe human society in its total environment as an overall system whose behavior can actually be simulated and studied. (Peccei 1977, 69)

It is worth highlighting this decision because it clearly links global concerns with the need for global computer models. There is in this statement a conscious call for new ways of perceiving and thinking about the world. It is a specific instance of how discourses of globality call forth the need for new practices of inscription, calculation, and conceptualization. It is at this level, of linking new discursive directions with the awareness of a need for new practices of perception vis-à-vis the world, that new forms of governmentality emerge. When we look at these developments at the capillary level of new narratives, often emerging in mundane meetings, we can see that the world is changed, sometimes dramatically, not by bold actions, but by minute transformations in discourse and practice.

The first plan proposed to the group was produced by Ozbekhan. The most important problem that the Club of Rome had, and which Ozbekhan tried to solve, was conceptualizing quantitatively and mathematically the many subsystems and interrelations that constituted the global system. This conceptualization had to be sufficiently complex to have scientific validity, sufficiently flexible to describe the "world problematique"—the interrelationship of world problems—and sufficiently unencumbered by theory to be attractive to potential financial backers. The Club of Rome itself, run largely on a shoestring budget, had no funding to support the type of research envisioned.[1]

After some months of work, Ozbekhan presented his plan entitled "Quest for Structured Responses to Growing World-wide Complexities and Uncertainties." It consisted of a fifty-nine page proposal with a series of appendixes adding another sixteen pages.[2] The proposal was open-ended in that the research into the constructions of the models and the information to be fed into them would produce results that could not be anticipated. Furthermore, the models would, in turn, be integrated into one final model which was to produce specific policy suggestions.

Ozbekhan started his proposal by pointing out the rapid world-wide acceleration of change. His proposal, like other discourses about globality, was structured around notions of speed—of careening toward immense disasters at high velocity. The very condition of speed is used in support of the argument that the human mind by itself can no longer cope with the speed and complexity and, therefore, needs to be augmented or even partially supplanted by new tools, such as anticipatory modeling. Ozbekhan argued that the problems of the present (his proposal was written in 1970) could not be handled by the normal tools available to governing officials and corporate managers. He suggested that today's problems have several new characteristics which make "the normal tools" obsolete. These characteristics included the global scope of the problems; their interactivity, sometimes obscure and difficult to spot; and the accelerating rate of change.[3]

In order to conceptualize the new complex situation as he saw it and as the Club of Rome wanted to see it, Ozbekhan developed the notion of "problematique," a term that the Club of Rome has since used in their publications long after Ozbekhan broke his connections with them. Peccei often talked of the "world problematique" as the area of concern for the Club of Rome. In his original proposal Ozbekhan introduced the notion of problematique by arguing that

> when we consider the truly critical issues of our time such as environmental deterioration, poverty, endemic ill-health, urban blight,

criminality, etc., we find it virtually impossible to view them as problems that exist in isolation—or, as problems capable of being solved *in their own terms*. For even the most cursory examination will at least reveal the more obvious (though not necessarily the most important) links between problems. Where endemic ill-health exists, poverty cannot easily be divorced from it, or *vice versa*. Certain kinds of criminal behavior often, though not always, seem to be related to poverty or slum living conditions. Furthermore, if we try to solve any such problems exclusively in their own terms we quickly discover that what we take to be the solution of one category of problems may itself generate problems of another category (the reduction in death rate in developing areas and the resultant increase in poverty, public unrest, overpopulation, etc., is a good example of this single avenue approach) . . . it is this generalized meta-problem (or *meta-system of problems*) which we have called and shall continue to call the "problematique" that inheres in our situation.[4]

After defining the notion of "problematique," Ozbekhan proceeded to list what he now called the "Continuous Critical Problems" that were part and parcel of the problematique. As mentioned above, in his talk at the OECD Bellagio Conference Ozbekhan had listed twenty-eight such problems. Now a few months later his list had grown to forty-seven. These problems included: "generalized lack of agreed-on alternatives to present trends," "widespread failure to stimulate man's creative capacity to confront the future," "limited understanding of what is 'feasible' in the way of corrective measures," "fast obsolescing political structures and processes," "unimaginative conceptions of world-order and of the rule of law," and "insufficient understanding of Continuous Critical Problems."[5] The list included many other problems that we would easily expect to be included in such lists, such as ecological degradation, widespread poverty, and population growth. These problems were presented as interrelated and thus, both implicitly and explicitly, call for a new way of seeing and organizing the world—a way that, not coincidentally, would give systemic planners such as Ozbekhan and many members of the Club of Rome plenty of employment. The discursive strategies of this proposal, as in many other discourses of globality, included the generation of arguments for new spaces of governmentality and new institutions and new techniques for studying and manipulating the new spaces. This is not to suggest, uncharitably, that Ozbekhan or the Club of Rome were doing no more than feathering their beds. Rather, their discourses of globality— in tying together different problems, in calling for new ways of thinking— necessarily lend support to new practices, such as computer modeling. The new practices gain their very legitimacy from the stated need to conceptualize in new ways. Thus, the discourses and practices of globality, together with their institutional structures, can be said to be mutually constitutive. It is this aspect that lends them an air of "obviousness" and ultimately of "normality" (in the sense that it is "normal" to think that our problems are obviously "global").

Ozbekhan's proposal was permeated with the language of cybernetics and systems thinking. Thus, for example, in talking about the Continuous Critical Problems, he states:

> Neither their rate of occurrence nor their intensity is uniform throughout the world. Therefore, the causality structure that underlies such a listing is obviously of extreme complexity and actually impossible fully to ascertain through mere observation. . . . These large problems are system-wide, interdependent, interactive and intersensitive, they transcend national frontiers, or even regional boundaries; and . . . they are seemingly immune to linear or sequential resolution. . . . In actual fact the situation tends increasingly to appear as a single complex system whose internal relationships, interactions, fields of force, and overlaps are extremely confused (Fig. 4) and impossible to delineate without a very serious attempt to model it in its entirety. Such a modeling effort could, for example, reveal the morphology of the situation as resembling what is shown in Fig. 5—namely, as having a composite dynamic core, and differing intensities of interfaces and relationships, all of which must be identified and organized into a unified frame of perception and understanding. (Ozbekhan 1970, 17–19)[6]

This cybernetics language "naturally" calls for cybernetic solutions to grasp the structures of the systems under consideration. Again, the discursive strategy calls forth the type of practice, such as computer modeling, that lends authority to the discursive conceptualization.

Ozbekhan went on to argue in this proposal that the modeling he called for needed to be guided by some underlying "value-base." As he stated:

> The primary aim of modeling is to give the subject a shape, a structure, a configuration that is determined by an objective which, itself, is external to the subject. Hence the clarifications or insights that might be obtained from a successful modeling effort are never reached in terms of the subject (i.e., a problem or situation) but in terms of the external objective to satisfy with which the modeling was undertaken in the first place. Such an objective always entails a value, and the setting of it must therefore create the particular *value-base* that gives meaning and direction to the whole endeavor. (Ozbekhan 1970, 23)

In order to elucidate a possible value-base, he looked at the list of Continuous Critical Problems and concluded that they all exist in relation to something else, such as other problems or embedded cultural values that are taken for granted. He concluded that we experience problems due to some *imbalance* between "situational components" of the system. He drew a connection with notions in ecology, where "imbalance" defines the pathology of any given ecosystem. From this connection he then proposed that the value-base needed for the proposed research rests on notions of ecological balance. He stated:

> Hence if we extend, as is increasingly being done nowadays, the definition of ecology to comprise all the dimensions of occurrence in our world-wide environments it becomes possible to say that we are confronted with a problematique which is ecosystemic in character. The normative statement that describes the value-content of any ecosystem is "ecological balance." Consequently it is the idea of *ecological balance* that can, and will, be taken as the underlying value-base of the study; for in terms dictated by our situation the "good" is self-evidently and most generally capable of being defined as the re-establishment of that many-dimensional dynamic balance that seems to have been lost in the modern world. (Ozbekhan 1970, 26)

As suggested in the above passage, Ozbekhan's notion of system includes elements, such as values that come to the system from the outside, that are extrinsic to the system under study, and that must be explored in terms of externalities relating to the system rather than as internal elements of the system. In this conceptualization of system, he differs from Forrester, who tends to see values as emerging from within the system itself.

Ozbekhan's proposal included a cost estimate of $900,000 and a time frame of fifteen months.[7] He envisioned a research team composed of eleven senior scientists and an unspecified number of junior researchers from the areas of computer programming, logic systems, and data retrieval. The team of senior scientists was to include the following experts: three planners with cybernetics backgrounds, one mathematician, one senior statistician with an OR background, two senior computer programmers, two social scientists, one economist, and one political scientist with a background in international relations.[8] It is worth noting the interdisciplinary character of the proposed team, as well as its strong orientation toward what may be called, broadly speaking, "systems analysis." This proposed team was not dissimilar from those that emerged out of World War II around OR.[9] Ozbekhan expected that the work would be carried out in Europe, at the Battelle Institute in Geneva and with the use of Battelle's own computers or those of nearby Geneva University.[10] Ozbekhan also proposed having a steady stream of consultants from various disciplines, such as the life sciences, anthropology, education, and psychology, to provide the main research team with access to a wide diversity of views and knowledge practices.

While the proposal was seen as a very good effort by the Club of Rome executive committee, it did not meet with universal endorsement. Resistance emerged because the plan was seen as somewhat vague and too complex. It provided not a clear-cut tool with which to do research but rather a scheme to eventually get that tool. Both time and financing also emerged as serious issues. The Club of Rome wanted a working model within a year. Ozbekhan recommended a project lasting at least fifteen months and which made no firm commitment to producing the kind of powerful modeling tool the club was looking for. The open-ended nature of the proposal was not seen as providing the one-time big punch that the Club of Rome, especially Peccei, was looking for. Peccei described some of his doubts about the Ozbekhan proposal a few years later in his book *The Human Quality*. He wrote:

> What Ozbekhan had in mind was to develop an initial coarse-grained model, or models, of the world dynamic situation in the overall expectation that such models will reveal both those systemic components that are most critical and those interactions that are most generally dangerous for the future. . . . This approach was in principle reasonable. . . . What gave us most concern was whether the phase so outlined could be completed in a reasonable time even if it were possible to mobilize the best talents available. . . . The appeal of the project as a trail-blazer gradually faded as doubt about its feasibility grew. Building, for the first time, a descriptive model of the world is in itself a tremendous task; but the actual Programme was even more ambitious. It expected to incorporate in the model a series of goals, with a view to obtaining some preliminary indications of policy alternatives to attain them. We estimated that this would require such a great advance in the state of the art of systems analysis that nobody could predict whether and when the task could be accomplished. (Peccei 1977, 71)

Furthermore, the Volkswagen Foundation, to which an appeal for money had been made, turned the proposal down, on grounds of insufficient methodological focus (Pestel 1989).[11]

At this point members of the Club of Rome were losing confidence that a manageable model, embodying the club's concerns, could be built. It was within this atmosphere of uncertainty and lack of clear sense of future direction that the first general meeting of the Club of Rome took place on June 29, 1970, at Bern, Switzerland. The meeting was convened to discuss the research program around the notion of the world problematique and the interrelationship of all problems currently affecting the planet. The initial idea was to establish a research center in Bern itself, a sort of neutral place between East and West, with Ozbekhan as director.

The meeting soon became bogged down in trying to simplify Ozbekhan's plan and obtain funding for the project. Pestel had confirmed for everyone in attendance that the Volkswagen Foundation would not finance the project unless it was simplified (Pestel 1989). It was at that juncture that, as Pestel remembers it, "Professor Jay Forrester of MIT stepped in deus ex machina to announce that his method, then still known as 'Industrial Dynamics' could do the job" (Pestel 1989, 23). It was not quite that simple, though it probably felt that way to Pestel, who like many other members of the club had grown increasingly concerned about getting anything off the ground. At this point, as Forrester put it, the club had a project with "no methodology and no money, which is reasonably close to no project."[12]

The Bern Meeting of June 1970 and Jay Forrester

Jay Forrester had attended the Bern meeting at the invitation of Carroll Wilson, himself a new member of the Club of Rome, and with the blessings of Peccei. Peccei had invited Forrester to become a member of the Club of Rome in March of that year, but it was Wilson who encouraged Forrester to attend the club's first general meeting.[13] Carroll Wilson is an interesting figure in his own right. He himself was invited to become a member of the Club of Rome in February of 1970, most probably at the suggestion of Alexander King, with whom he was friendly.[14] Like Forrester, Wilson was connected to MIT for a large part of his life. Having received his B.S. from MIT in 1932, he started work as assistant to Karl T. Compton, then president of MIT.[15] He also worked closely with Vannevar Bush, who was then dean of engineering. In 1940 he followed Bush to Washington, DC, where Bush was to organize the NDRC as well as the successor to NDRC, the OSRD.

At the end of the war, Carroll Wilson went to serve as science advisor to the State Department on atomic energy issues, eventually becoming secretary of the Board of Advisors. In that capacity he helped draft the governmental *Report on the International Control of Atomic Energy*. In 1947

President Harry Truman named him as the first general manager of the Atomic Energy Commission, a post in which he served for three years. After serving in that position, Wilson went into the private sector to act as president and general manager of two companies involved in radioactive materials mining. In 1959, he returned to MIT to teach at the Sloan School of Management. He continued to be associated in an official or semiofficial capacity with MIT until his death in 1983. While at MIT, he was involved with groups within and outside MIT concerned with issues of foreign relations and Third World development. He established, for example, the MIT Fellows in Africa Program to train managers for newly independent African nations. He was also involved in international programs dealing with the production of biodegradable pesticides, scientific advice to Third World countries, environment, and energy issues. He organized the World Coal Study, and was involved in a two-year study of World Energy Prospects, 1985–2000. Like Forrester, who was his colleague of many years at the Sloan School of Management, Wilson typified the engineer-scientist who emerges from involvement in World War II and prepares to extend and adapt his technical knowledge to wider and wider areas of endeavor.

The Bern meeting in June of 1970 was Forrester's first formal activity with the Club of Rome to which he came more as observer than as an active participant.[16] By late afternoon, the meeting was mired in seemingly endless discussions of funding when Forrester finally spoke and suggested that his methodology of Industrial Dynamics could be helpful. He invited members of the club to come to Cambridge and see for themselves what he and his group could do. But, he added, "they would have to come for two weeks or not all."[17] This demand Forrester felt was essential because, in his experience with studies of the city using his methodology, it took that long for people to understand the counterintuitive nature of the problems they were looking at. At first this suggestion fell by the wayside. A bit later, after more fruitless discussion of the "no methodology, no money" problem, one of the participants, most possibly Pestel, suggested that they look at Forrester's offer more seriously.[18] At that point there was some discussion of Forrester's ideas, and it was decided that a special session be set up that evening, after dinner, to further discuss the Industrial Dynamics approach. This session took place from 10:00 P.M. until midnight. The executive committee finally agreed to go to MIT for two weeks, in three weeks from that day. As Forrester observes, "it was a measure of the degree of concern about the issues" that they agreed to cancel two weeks of their already full schedules in order to attend the Cambridge meeting.[19]

The next day Forrester called his staff from Bern to have them start planning the logistics of the meeting. Aboard the plane, on the way back from Europe, Forrester himself sketched a rough scheme of the modeling that he wanted to prepare for the visitors. Eventually, he would have ready a practical demonstration of the modeling which was to clinch the final selection of his methodology and conceptualization.

It may be useful to pause and recapitulate the needs of the Club of Rome as members understood them in June of 1970. Pestel's recollection suggests that the club had very clear ideas that were only waiting for the proper quantitative model. Historical records suggest that the situation was more complicated. The members of the club had certain goals, which included the need to "jump-start" a worldwide dialogue about the predicament of mankind. Pestel and Peccei argued that the best way to initiate such a dialogue was to model and quantify the predicament and, thus, present it in a fresh manner. The very process of going about finding such a model, however, was quite complicated and involved overcoming many obstacles, most especially funding and timeliness. This last aspect was a self-imposed stricture, but intrinsic to the process that would culminate in choosing Forrester's modeling techniques as the most appropriate to achieving the club's goals.

Furthermore, Peccei and others had some general ideas that, without a methodology, lacked clarity. The process of seeing their goals eventually embodied in a concise computer model was not the same as simply finding the right model, as going to the store to find the right curtains for a new house would be. The failure of the Ozbekhan proposal to become operational illustrated the fact that this was not a simple problem. It involved serious and complex negotiations that ultimately coalesced around a particular direction. Issues included the ways in which subsystems within the larger global system related to each other, what variables were important, and how the modeling would be made to produce meaningful results and yet remain somewhat manageable. The tool that emerged from discussions and negotiations necessitated a transformation of both the outlook of the Club of Rome and of Forrester's model. Much of this was done at MIT, as described below. In other words, Forrester did not appear in a space perfectly fit for his ideas. The process was one of mutual enrollment. Forrester's growing relationship with the Club of Rome and response to the needs of the Club of Rome constructed the space for collaboration as surely as did the club's goals. Funding and timing were not trivial aspects of the project either, ones that could be waived aside by the brilliant connection between the Club of Rome and Forrester. In fact, the eventual disillusionment with *Limits to Growth* on the part of several members of the Club of Rome, including Pestel, suggests that the chosen model was never comfortably stable, and issues of time and money might have played important roles in the ultimate selection of Forrester's ideas. It is also worth emphasizing that Forrester's ideas were not finalized when he attended the Bern meeting. In fact, as he put it to me in an interview, the very modeling of *World Dynamics* "grew rather directly out of the meeting in Bern."[20]

The historically contingent aspect of the whole endeavor must be stressed. Otherwise we fall into the trap of constructing a Whiggish narrative of how *Limits to Growth* came to be. In fact, people like Peccei and Pestel sometimes tend to argue along the lines that the existence of global problems logically called for a global model to make sense of them. The very notion of global problem, however, emerged historically out of precisely the type of practices that the Club of Rome and Forrester engaged in. The discourses about global problems and the practices that have embodied those discourses, as well as the fields of intervention that they brought into existence, emerged in a mutually constitutive and historically contingent fashion. The process was one of historical construction, not of a natural logic that apprehended already existing essences called "global problems."

A short detour into another project that occupied Peccei's energies, namely, the establishment of the IIASA, can serve to illustrate this point further. The detour will show that the global perspective for which the Club of Rome eventually became famous was not the logical end result of the questions that interested Peccei. In fact, Peccei started participating in the IIASA project with premises similar to those advocated in his writings about the Club of Rome. The project, however, ended by having a different character than that of the Club of Rome. Even though Peccei started with similar ideas, the process of enrollment and the diverse goals of the different actors involved in both projects culminated in fairly different organizations and perspectives.

The Establishment of the International Institute for Applied Systems Analysis

In 1966 while Peccei was giving a series of lectures on "the challenges of the 1970s for the world of today" in Washington, DC, he made contact with Hubert Humphrey, then vice-president in the Johnson administration. Apparently his ideas were warmly listened to, and, on the strength of that reception, Peccei wrote a memorandum to Humphrey arguing for the need to start a multinational project to look at international problems. A few months later, in December of that year, McGeorge Bundy, newly appointed president of the Ford Foundation, called a press conference to announce that President Johnson had named him his personal representative to explore "the possibility of establishing an international center for studies of the common problems of advanced societies" (Peccei 1977, 51). Several years of negotiations with representatives of sixteen countries culminated in October of 1972 with the establishment of the IIASA in Austria.

Peccei was involved all along in this process as a sort of "nonofficial" representative from Italy. He mediated at least twice between American and Soviet representatives when it seemed as if the whole project would become mired in superpower politics.

While the institute was largely the result of Peccei's initial lobbying, he was not totally satisfied with the end result. As he put it:

> All this time I was aware that through it [IIASA] and by means of it we could explore only some facets of reality or prepare to face up only partially to its bewildering complexity. Other research and other insights were also necessary if we were going to understand mankind's fundamental problems. (Peccei 1977, 53)

One thing that possibly bothered Peccei about the IIASA setup was the lack of a global perspective in its mandate, a perspective to which he was increasingly committed. It is important to point out that, when Peccei first started lobbying for an international organization, he was really interested in an institution that would explore possible responses from the West to China. Peccei saw China as looming increasingly large and possibly dangerous in its posture toward the West. By the time the IIASA was set up, Peccei had moved away from this concern and had started to advocate the kind of global perspectives that would accompany all of his activities until his untimely death in 1984. In a certain way the IIASA became the kind of institution he had foreseen in the late 1960s, but the actualization of his vision in 1972 was no longer satisfactory, and the IIASA ended up being something of a disappointment to him.

The IIASA's original mandate was to study the seemingly immense problems of Western societies. The major objective of the IIASA, as embodied in its original charter, was to "initiate and support collaborative and individual research in relation to problems of modern societies, arising from scientific and technological development."[21] When the institute finally started functioning in 1972, its research projects were focused on problems of industrial societies. Interestingly, however, by 1977 the goals had shifted toward more global problems and global modeling. In fact, an in-depth analysis of the first two projects of the Club of Rome, *Limits to Growth* and *Mankind at the Turning Point* (Mesarovic and Pestel 1974), was the catalyst in the shift of perspective at the institute (Pauli 1987).[22]

Illustrative of the gathering strength of the discourses and practices of globality is the further development of global change studies at the IIASA. The shifting focus from industrial societies' problems to global problems underwent further transformation such that by 1982 it became a program on the "Sustainable Development of the Biosphere."[23] That program, eventually called "Ecologically Sustainable Development of the Biosphere," was initiated in 1982 when C. S. Holling, then director of the IIASA, contacted William C. Clark, a scientist at the Institute for Energy Analysis in Oak Ridge, Tennessee. Clark was asked to outline a research program that would "most help to realize the potential for better management of interactions between development and environment." After reviewing many programs then under way, Clark and his team decided that their program "would best complement existing work on environment-development interactions by emphasizing four characteristics: a synoptic perspective, a long-term time horizon, a regional to global scale, and a management orientation" (Clark 1986, 7).[24]

It is worth stressing the fourth characteristic of the IIASA program—the focus on management. In previous chapters I talked about the way in which the discourses and practices of globality come together and expand. The IIASA program can be seen as another instance of the tendency of scientific explorations of the planet to get linked with specific policy issues and, in turn, demand processes of management. In the introductory chapter to the published collection of papers that eventually came out of the IIASA program, Clark calls for the application of game theory modeling, among other strategies, to look at the issues involved in planet management (Clark and Munn 1986). In so doing, Clark adapted and extended some of the techniques developed during World War II and linked them with discourses of globality which culminated in developing processes of planet management.

The Pivotal Meeting: Cambridge, Massachusetts, 1970

While the long process that culminated in the establishment of the IIASA was going on, plans for the Club of Rome projects were also moving ahead. On June 29, 1970, at a meeting of the Club of Rome in Bern, Forrester made an initial presentation of his Industrial Dynamics program and asserted that it could best serve the needs of the club. The members present at that meeting were suitably impressed and gave Forrester provisional approval to start the construction of a working model of the interactions of Earth and human activity using the concepts of Industrial Dynamics, later to be renamed System Dynamics. Funding for this provisional project was obtained from the Volkswagen Foundation (Pestel 1989).

Immediately after returning from the Club of Rome meeting in Bern, Forrester got to work at a feverish pace for a more formal presentation of his theories and modeling to three representatives of the Club of Rome executive committee. At the Bern meeting Forrester was tentatively successful in convincing several of the club members of the feasibility of his methodology and of the fit of his modeling with the club's goals. Now in Cambridge he had to solidify those initial impressions and convince the club, in the persons of Peccei, Pestel, and Thiemann, to commit to his methods. In addition to the three members of the executive committee of the club, Ozbekhan and Jantsch were also part of the visiting group.

The meeting that Forrester and his associates put together was impressive. The whole affair lasted twelve days, from July 20 to July 31. Most days were packed with talks, activities, and learning interactions. The first day, for example, was programmed with activities from 9:00 A.M. to 5:00 P.M. The program for that day added that the "evening" was "free" but suggested that the time could be used for "reading."[25]

The conference, as conceived by the MIT team, had a number of different functions, including teaching the basic concepts of System Dynamics to visiting club members and providing the club with enough information to allow them to decide what type of formal relation with the MIT team they might want to establish. Another unstated but present function or goal of the seminar was to determine how to sort out the relations between the MIT team and Ozbekhan and his group in Europe. A final goal was to allow Peccei to educate Forrester's team on the overarching goals of the Club of Rome. All of these goals were eventually accomplished during those twelve days.

The backbone of the seminar Forrester prepared for the club had been used before to train corporate managers in the use of System Dynamics concepts. However, since the Club of Rome was interested in global problems, Forrester did not want to focus on the usual corporate examples and a refined World1 modeling program was built instead.[26]

The seminar began with general remarks by Forrester followed by an explanation of the Club of Rome objectives by Peccei. Forrester stressed the short amount of time that he had to pull things together since his initial meeting with the Club of Rome members in Bern at the end of June. He also stressed his confidence that the social system modeling that he developed could meet the needs of the Club of Rome.[27]

Peccei's presentation briefly described the nature of the Club of Rome. It also stated the basic assumptions that the club had in terms

of awakening interest in the study of the "predicament of mankind."
He pointed out that in the club's estimation the time available to avoid
"monumental crisis" was measured in years, not decades. He stressed the
club's understanding that some type of systemic worldwide modeling was
the best way of approaching the study of problems on a planetary scale.
The main goal of the three representatives of the executive committee
of the Club of Rome at MIT was to learn "about work on structure and
behavior of large-scale systems." They also sought to learn about the types
of trade-off that would need to be made between the different, as yet
undefined, subsystems composing a worldwide system. He felt the main
contribution from the MIT team would be in the area of methodology.[28]

The last important presentation during the first day of the conference was given by Gordon Brown, who had been, for almost ten years,
dean of the School of Engineering at MIT and was, at the time of the Club
of Rome conference, a chaired professor of engineering. Brown was and
remained a pivotal mentor in the career of Forrester. His talk to the
conference was a historical presentation, focusing on Forrester's career
and the long and successful history of systems research at MIT.[29] This presentation no doubt provided added weight to Forrester's assertions that
he was the ideal person for the job. During the next nine days, many other
presentations were given, covering applications of Industrial Dynamics to
the corporation, the city, community health care, drug addiction, and the
modeling of Malthus's ideas. This last presentation was given by Gerald
Barney. He had received a doctoral degree in industrial dynamics and
was working at the Center for Naval Analysis as a postdoctoral fellow.[30]
Barney eventually became study director of the *Global 2000 Report to the
President* (Barney 1981). This report was initiated under a directive from
then President Jimmy Carter who was strongly attracted to the promise
of global modeling and its assumptions about the interdependence of
world problems (Richardson 1987).

Other presentations during the conference were given by Dennis
Meadows, who would eventually become the leader of the *Limits to Growth*
project.[31] In 1969 Meadows had received a Ph.D. in System Dynamics from
MIT's Sloan School of Management. That same year he was hired as an
assistant professor at MIT. Meadows was on leave that first year, having
gone with his wife, Donella, on a trip, driving from Europe to Sri Lanka.[32]
The trip was cut short by three months, and the Meadows were back in
Cambridge the day after Forrester had returned from Bern. Forrester
asked Meadows to participate in the meeting with the Club of Rome and
he accepted. As Meadows later recollected in an interview, this offer came
at a particularly fortuitous time since, having been gone for almost a year,

Meadows had no backlog of work to get back to and could dive into the project full-time.[33]

One of the climactic moments of the conference came on the third day when Forrester ran the model by simulating different assumptions about the impact of technology to alleviate expected shortages of resources and other environmental problems. It was a dramatic moment, according to Forrester, and was to have lasting influence on the Club of Rome members, especially Peccei.[34] As Pestel recollects:

> I still remember how impressed Aurelio Peccei was by the fact that all computer runs exhibited—sooner or later at some point in time during the next century—a collapse mode regardless of any "technological fixes" employed. Peccei obviously saw his fears confirmed that the continuation of exponential industrial and population growth would eventually result in disaster at some time in the not-too-distant future. (Pestel 1989, 24)

As the members of the executive committee of the club became enthusiastic about the potential of computer modeling based on System Dynamics, both Dennis and Donella Meadows began to feel that they would like to be further involved. To that end, Dennis Meadows presented a two-page memorandum to the Club of Rome detailing how the modeling could be accomplished under his direction and including an estimated budget.[35] The club warmed up to the idea of the project, while Ozbekhan, who was offered the possibility of collaborating in some capacity with an MIT team, decided instead to leave the project and withdraw his membership in the Club of Rome altogether. According to Ozbekhan, he left the project because he felt that the modeling presented by Forrester and his team was too superficial.[36] According to Dennis Meadows, the break came because of a gulf in methodological approaches between Ozbekhan and the MIT team and questions about overall leadership of the project. Furthermore, according to Meadows, Ozbekhan's methods had already been rejected by the Volkswagen Foundation, so the club needed a different approach.[37]

The Club of Rome eventually gave the go-ahead for the MIT team to prepare a proposal for a full-fledged study. This study would involve the grounding of Forrester's "World2" in the empirical literature and further elaboration of the model's structure.[38] The process of finding empirical data to run the model, something which Forrester's original model lacked and for which he was much criticized, was also fairly problematic for Meadows and his team. As Dennis Meadows put it, "it was hard in those days to get the kind of comprehensive, cross-national time-series data on the issues we wanted to see, except on population. So we were looking where we could, with the United Nations and the World Bank as principal sources of information."[39]

The issue of available information is worth highlighting because it illustrates the close connection that exists between viewpoint and the information collected to support that viewpoint. In terms of the paradigm shift of forms of governmentality—from one centered on the nation-state to one centered on the planet itself—the very viewpoint determines the type of information gathered and the type of practices used to gather that information which, in turn, undergird the specific form of governmentality from which they emerge. Thus, the Meadows team had problems finding globally oriented information because information of that type was not normally collected within a framework of scientific practice that did not conceptualize the global as an entity on which information needed to be collected. In order to meet their modeling needs, they had to improvise and aggregate whatever information they were able to gather that was of an international nature. Information on population, for example, was readily available. This availability is readily explainable by the fact that the concept of population had been central to the form of governmentality centered on the Enlightenment notion of "man" (population being "man" in the aggregate). In other words, the Meadows team could easily find information on population because the discourses and practices constitutive of the older governmentality conceived of population as a necessary object of research. The notion of population has been carried forward within the emerging forms of planetary governmentality but with changing emphasis. The older emphasis was on population as that which needed to be managed largely within the borders of nation-states. The emerging emphasis sees population as a natural force, measured in terms of anthropogenic effects, such as pollution and ozone depletion, on the environment. As reconstructed within a governmentality centered on the planet as a whole, the meaning of "population" derives from the larger context of the planet itself.[40]

Once the members of the Club of Rome left Cambridge, having accepted the Meadows plan, Meadows and his team had the immediate goal of securing funding. The Volkswagen Foundation was to be approached again in time for their November board meeting. By mid-August the MIT team had a draft of the proposal. The proposal dealt with phase 1 of a two-phase project. Phase 1 would consist of a modeling project done at MIT. Eventually, the plan was to transfer the project to Geneva for further modeling as phase 2. Phase 2 never came to fruition.[41]

Notes

1. Running the club meetings would become increasingly expensive. Particularly expensive were the meetings of the whole membership. In order to cover these costs, the club became increasingly dependent on the generosity of host governments, sometimes with deleterious results. Some hosts wanted to get their money's worth in terms of political propaganda, and thus members of the club found themselves spending more and more time listening to the canned speeches of politicians from the host country. This certainly was the perception that Dennis Meadows had of later meetings (Dennis Meadows, interview by the author, May 31, 1994). It was also one of the issues that eventually led Carroll Wilson to resign from the Club of Rome. See note 11 below.

2. A copy of the proposal can be found in the Carroll Louis Wilson Papers, MC29-55-2113. The following discussion, including references to page numbers, is based on this copy. An outline of the work program based on this proposal, providing some rough estimates of work schedules, can be found in the Gordon Stanley Brown Papers, MC24-17-686.

3. Carroll Louis Wilson Papers, MC29-55-2113, p. 4.

4. Ibid., pp. 12–13.

5. Ibid., pp. 14–17.

6. Figure 4 shows geometric figures, such as triangles and circles, representing different critical problems, all tightly intertwined. Figure 5 shows the same tight mix of geometric figures as does Figure 4 but darkens in the core overlapping areas of the figures, suggesting an internal structure not perceivable at the surface.

7. Information on cost and duration is described on page 35 of the proposal (Carroll Louis Wilson Papers, MC29-55-2113).

8. Ibid., p. 75.

9. I discussed these early OR teams in chapter 2.

10. Carroll Louis Wilson Papers, MC29-55-2113, p. 75.

11. The approach to the Volkswagen Foundation was made by Pestel, then a member of the board of directors of the foundation. Reliance on corporate support was typical of the operations of the Club of Rome. The support was not always forthcoming since some corporations were cool to the Club of Rome's projects. Peccei, by and large, resisted government support to avoid the appearance of taking political sides. There were exceptions, as money was given by some

"politically safe" European countries (e.g., Sweden, Austria). A conference in Mexico, financed in part by the Echeverría government, became an example of the problems attached to government financing. The Mexico conference was a fiasco and led, among other things, to Carroll Wilson's resignation from the Club of Rome. Corporate financing was not, of course, a sign of neutrality, but was preferred. It should be pointed out that the Club of Rome operated on a very thin purse, partly by design (to avoid empire building) but partly by necessity. Not many companies were willing to support an organization identified with "no-growth" ideas and policies.

12. Forrester, interview by the author, December 12, 1991.

13. Peccei's letter to Forrester is dated March 4, 1970. A copy can be found in the Carroll Louis Wilson Papers, MC29-54-2092.

14. Ibid. A letter from Peccei to Carroll Wilson dated March 4 acknowledges a February 23 letter from Wilson to Peccei accepting membership in the Club of Rome. Peccei, ever alert to leveraging all personal relationships in order to connect with more people or to deepen a relationship, asked Wilson's advice on how to best invite the Russian Gvishiani to the upcoming Club of Rome meeting in Bern. Wilson and Gvishiani had developed a cordial working relationship in international scientific meetings.

15. The biographical information on Carroll Wilson is included in the "finding document" for the Carroll Wilson Papers at the MIT archives. The information (manuscript collection MC29) was compiled by Amy G. Sugerman. Further information is gathered from volume 2 of the history of the Atomic Energy Commission by Richard Hewlett and Francis Duncan (Hewlett and Duncan 1972).

16. Forrester, interview by the author, December 12, 1991.

17. Ibid.

18. Forrester did not remember with certainty, though he thought it was Pestel. In his recollections Pestel does not address this particular point but states that Forrester's "straightforward approach was particularly appealing" to him (Pestel 1989, 23).

19. Forrester, interview by the author, December 12, 1991.

20. Ibid.

21. Quoted in a statement by Peter Janosi of the American Academy of Arts and Sciences during the congressional hearings on American participation in funding the IIASA (U.S. House Committee on Science, Space, and Technology 1990, 18).

22. The IIASA was the first large scientific collaboration between the United States and the then Soviet Union and had resulted from a thaw in the cold war with the advent of détente. Its fate was tied fairly closely to the fate of détente. During the Reagan administration, a toughening of the American posture vis-à-vis the Soviet Union resulted in U.S. withdrawal from active membership in the IIASA, dealing the institute a large financial and scientific blow. Members of the Reagan administration had argued that the IIASA was serving to transfer sensitive computer technology to the Soviet Union. There had also been a

case of a Soviet spy using his position in the IIASA for espionage purposes. Eventually, with the collapse of the Soviet Union, American membership in the IIASA was reestablished. Among the main arguments for the need of American participation were: aiding Eastern European nations with the transition from a centrally planned to a free market economy, and the unique position of the IIASA as an international organization in doing research on issues of global change and global sustainable economics. Hearings on U.S. participation in the IIASA were held on April 18, 1990. For further information see the U.S. House Committee on Science, Space, and Technology (1990).

23. The transformations along the way included studies on the circulation of carbon dioxide in the atmosphere and the impact of climatic change variations on agriculture. The computer modeling that the IIASA produced between 1976 and 1985 from within their Food and Agriculture Program served as the basis of the U.S. Environmental Protection Agency's 1989 study on "Policy Options for Stabilizing Global Climate" (U.S. House Committee on Science, Space, and Technology 1990).

24. Included among those programs were UNESCO's "Man and the Biosphere," NASA's "Global Habitability Program," the World Meteorological Organization's "World Climate Programme," the OECD's "Economic and Ecological Independence," and the World Resource Institute's "The Global Possible." It is worth noting that this listing is only partial, but it serves to illustrate the veritable avalanche of studies that posit the whole planet as a field of study, concern, intervention, and management.

25. Program, Club of Rome Conference, MIT. A copy of this program can be found in the Gordon Stanley Brown Papers, MC24-17-684. Most days were filled with talks by different members of the System Dynamics group at MIT. Forrester dominated the conference with several, almost day-long, presentations.

26. Dennis Meadows, interview by the author, May 31, 1994.

27. The content of Forrester's presentation can be found in summarized form in the Gordon Stanley Brown Papers, MC24-17-684.

28. Ibid.

29. I discuss the nature and importance of the relationship between Forrester and Brown in chapter 3 above ("From Servomechanisms to Planet Management"). Chapter 1 also traces the history of the emergence of the technoscientific practices and discourses that have animated "systems" and have facilitated the conceptualizations of Earth and human activity as a total, irreducible system. World War II and its aftermath are pivotal in those developments.

30. For a complete list of staff members associated with the conference see the Gordon Stanley Brown Papers, MC24-17-684. This file has a list of all attendees as well.

31. One of his presentations was on the fluctuations of commodity prices and their relation to production, consumption, and pricing. This was the focus of his doctoral dissertation at MIT. A copy of this talk can be found in the Gordon Stanley Brown Papers, MC24-17-684.

32. Meadows, interview by the author May 31, 1994.

33. Ibid.

34. Forrester, interview by the author, December 12, 1991.

35. Meadows, interview by the author, May 31, 1994.

36. Hasan Ozbekhan, interview by the author, May 27, 1994.

37. Meadows, interview by the author, May 31, 1994.

38. Ibid.

39. Ibid.

40. It is outside the purview of the present work to explore this issue in great detail. But an exploration of the changing notion of population would add meaningfully to the genealogy of planet management. I think such research would highlight the discontinuity of discourses and practices that the employment of the same term "population" now obscures. Moreover, such research would uncover a move away from "population growth" as a threat to humans per se toward the conceptualization of population as a "natural" planetary force.

41. Meadows, interview by the author, May 31, 1994. Phase 2 was originally conceived as a polycentric project, with a steering and coordinating committee in Geneva and a network of interrelated research, with different methodologies and assumptions, carried out in different parts of the world. A description of phase 2 can be found in the convocation and annotated agenda for a general meeting of the Club of Rome in Canada in April of 1971. See the Carroll Louis Wilson Papers, MC29-54-2093, for a copy of the convocation and agenda. An earlier description can also be found in a letter that Thiemann wrote to Forrester. See the Gordon Stanley Brown Papers, MC24-17-684.

Six

The Limits to Growth

Originally, the Club of Rome was under the impression that Forrester himself would direct the project. This was not Forrester's understanding. He had already committed himself to doing research on a model of the national economy, with funding from the Ford Foundation, and did not want to take charge of the new project.[1] He remained formally connected to the project in an advisory capacity. He was paid a $7,000 consulting fee for the period from November 16, 1970, to August 31, 1971.[2] The leadership of the project was won by Dennis Meadows. He was fully trained in System Dynamics methodology, had the confidence of Forrester, and was one of very few people around who could take on a massive new project without much advance notice.[3]

As leader of the project, it was incumbent upon Meadows to organize the research team. The very structure of the project facilitated this task. As Meadows explained it in an interview, the planned computer model had five sectors—population growth, accelerating industrialization, environmental deterioration, malnutrition, and depletion of resources. So Meadows looked for people with expertise in those particular areas. The source of possible team members was limited to the Cambridge area, mainly graduate students from Harvard and MIT, as well as a couple of young German academicians. It was, as Meadows described it, "a motley crew."[4] The lack of more seasoned research team members is explained in large part by the sudden nature of the project and the desire, particularly by the Club of Rome, to finish the project within twelve to eighteen months.[5] It would have been rather hard to assemble a group of more established researchers on a short notice, since most of them would have been engaged with other projects.

The Meadows proposal, used to lobby the Volkswagen Foundation for financial support, opens with a short introduction that is packed with the language of speed and inevitability. Its first paragraph states:

> For some four thousand years the condition of the human race has been characterized by growth and change. Technological development has accelerated. Natural resources have been depleted. Our environment has been polluted at an ever-increasing rate. Population has multiplied at least 50-fold and may double again this century. Now there is evidence that change is occurring too quickly to permit adaptation by the planet's social institutions and its ecological systems. (Meadows 1970–71, 1)

This is certainly a hard-hitting paragraph and one which touches many issues found in discourses of globality, such as depletion and deterioration of the environment, concern with population growth, and the inability of current political and social structures to deal with the problems. The "systems" language of the paragraph reveals its conceptual framework.

The beginning of the second paragraph is equally pointed: "Growth cannot continue indefinitely on a finite planet." Meadows and his team go on to say that "we are faced with an inevitable transition from world-wide growth to global ecological equilibrium." They end this second paragraph by presenting the stark choices ahead:

> Because of the time delays inherent in social system change, decisions made now are already influencing the nature of that future equilibrium. Will it be an equilibrium of poisoned lakes, of oppressive crowding, of food shortage and a declining standard of living? Or will we choose a different mode of equilibrium characterized by a more desirable set of conditions? The shift from growth to dynamic balance may be initiated by a catastrophe such as war or starvation. Alternatively, transition could result from an enlightened, concerted, international effort to adopt new values and define new goals. (Meadows 1970–71, 1)

The introduction goes on to point out a lack in correct governance: while experts within their own fields can see the looming problems, they are at a loss to see these problems in the larger global context. They fail to perceive how these problems affect and are affected by other problems outside their immediate areas of expertise. Such limited perception may lead experts to propose solutions that may make sense in terms of their own area but that may be deleterious in terms of the whole. The authors of the proposal state:

> Demographers press for effective birth control measures. Ecologists seek an end to the destruction of our natural environment. Agricultural experts search for more efficient food production. All would admit that there are important interactions among their various approaches,

but the conceptual framework, the analytical methodologies and the vocabulary to unite the different fields, have been lacking. (Meadows 1970–71, 2)

The above quotation goes to the heart of the proposal, namely, the presentation of a method, System Dynamics, that will allow experts in the different fields to see the whole. The methodology of computer modeling, based on principles of System Dynamics, is squarely presented throughout the rest of the proposal as a tool that not only shows the interrelationships of the Continuous Critical Problems but that also enables experts from different areas to develop a common language and a common analytical approach. In fact, piggybacking on Forrester's claims, Meadows and his team make a move to suggest that System Dynamics can serve as a sort of universal language through which people from different disciplines can communicate.[6] Thus, on page 11 of the proposal, they write:

> One important advantage of System Dynamics for the Club of Rome
> program is that it represents real world relationships pictorially or
> mathematically in terms quickly learned by everyone. Sophisticated
> mathematical ability is *not* a prerequisite for understanding and using
> the results of a System Dynamics study. Thus demographers, economists,
> governmental leaders and others interested in global problems will be
> able to apply the Phase One [of CoR's Modeling] readily to their own
> fields. (Meadows 1970–71, 11)

The bulk of the narrative part of the proposal is dedicated to a discussion of the advantages of System Dynamics, with the exception of one-and-a-half pages dedicated to an overview of the Club of Rome. Among the Club of Rome's strengths are listed its "flexibility," since its membership is "being deliberately extended to include representatives of all cultures, and it is expected that the funds for its projects will be derived from organizations and individuals in several different nations" (Meadows 1970–71, 3).

The proposal gives a formal definition of System Dynamics as "a theory of system structure and a set of tools for representing complex systems and analyzing their dynamic behavior." A discussion of System Dynamics is prefaced by a history of the field, with the study of servo-mechanisms at MIT during the early 1940s credited as the birth site of the theory. In addition the authors cite the four historical streams of System Dynamics that Forrester suggested in his *Industrial Dynamics* (Forrester 1961). These are: (*a*) the theory of feedback control, (*b*) research on human decision-making processes, (c) the experimental model approach to complex systems, and (d) the digital computer, which permitted the simulation of realistic mathematical models.[7]

The writers of the proposal state that the insights from the early servomechanism research still inform the overall understanding of System Dynamics:

> Study of mechanical servo-mechanisms at MIT led to an awareness in the early 1940's that time delays, amplifications and structural relationships among a system's elements could be more important in determining aggregate system behavior than the individual components themselves. The concepts of information feedback and control were developed to express the relation between structure and behavior. More recent efforts to design automatic self-regulating control systems have extended these concepts and shown them to underlie behavior in all systems. (Meadows 1970–71, 7–8)

The proposal goes on to state that a fundamental advantage of System Dynamics is its focus on computer simulation, which is defined as "the process of conducting experiments on a model instead of attempting the experiments with the real system" (Meadows 1970–71, 8).

While stressing the value of simulation, the writers go to some lengths to state that simulation is not to be used to predict the future but rather to understand how different changes in a system may alter the behavior of that system over time and that eventually simulation may be used to develop the knowledge to manage the system. In terms of the global system, they claim:

> One can imagine many states which could characterize the globe several decades from now: international warfare, rampant epidemics or prosperous tranquillity. A System Dynamics study would be less useful in predicting which will exist than in indicating how alternative international agreements would alter the tendency to move toward each of those conditions. (Meadows 1970–71, 9)

In order to buttress the actual importance of simulation, the authors include summaries of three computer "runs" or simulations using the World2 model (the one that Forrester constructed for the Cambridge meeting discussed above). One of the runs is called the "standard run," provided to serve as reference. It shows a peak of population growth in the year 2030, followed by a gradual decline, which the authors attribute to a coming decline in natural resources. They also point out that in this "run" quality of life reaches its optimum point in 1970. In discussing this conclusion, they raise and respond to an evaluative question:

> Can this be reasonable considering today's strong worldwide feeling of distress and disenchantment? Perhaps so. A sense of well-being may be related more to "progress" and to improvements since the recallable past than to the absolute level of quality of life. A feeling of malaise could therefore occur at the peak of the quality of life curve because little improvement has been observed in the preceding two decades. (Meadows 1970–71, 18)

The next two runs included in the proposal have two distinct assumptions. One of them assumes that technology will improve to the point where the use of natural resources will be four times as efficient as in the standard run. The other run includes the first assumption and adds the variable that pollution will be reduced by 50 percent and capital investment in social priorities will increase by 20 percent. The whole exercise is included in the proposal primarily to illustrate the approach, not to predict outcomes.

The proposal included a budget for a total of $200,000 (the figure was later increased by $50,000) to cover the period from November 16, 1970, to August 31, 1971, when the project was expected to be finished.[8] The ability of the Meadows team to come up with a budget that was almost a quarter of the one that Ozbekhan had proposed no doubt was an important element in getting funding from the Volkswagen Foundation. In a cover letter that Pestel sent to the foundation, accompanying the new Meadows proposal, he highlights the vastly reduced cost of the project and the fact that two German scientists would be involved in the project.[9] In addition to the proposal, Pestel included a copy of the appendixes of papers that were distributed by Forrester during the Cambridge meeting. These appendixes were essentially a collection of examples on the application of System Dynamics to different areas of endeavor, such as commodity markets and drug addiction.

The Volkswagen Foundation was not easily persuaded to fund the new project. While it had given some financial support to the preliminary effort of "World1," it displayed much less enthusiasm for financially supporting the longer, more complicated, and more expensive project headed by Meadows. According to Pestel and Forrester, there was a fair amount of resistance to the idea of financing a project outside Europe (Pestel 1989). Originally, the research was going to be done in Bern, but with Forrester's involvement, the location was changed to Cambridge, Massachusetts. Heavy lobbying of the Volkswagen Foundation board members by Pestel, himself a member of the foundation, eventually assured financial support. A sum of $250,000 was committed to the project, marking the first time that money from this particular foundation crossed the Atlantic. For the Volkswagen Foundation and its activities, the *Limits to Growth* project was a watershed of sorts (Peccei 1977). According to Pestel, the early sentiment of the board members at the November 1970 meeting, when the funding decision was ultimately made, was decidedly against funding. As he put it:

> The board meeting . . . was very tense—at times even hectic. In the beginning of the long debate—to fund or not to fund—the prospects looked very bleak indeed. Gradually, the tide of opinion among the board members was turned around, and finally there was a large majority for the project. (Pestel 1989, 25)

With the funding assured, the final go-ahead was given.

A Recapitulation

In the early part of 1970, the Club of Rome was gearing up to organize and implement a research proposal on the world problematique being prepared by Ozbekhan. At the end of June of that year, coinciding with the first general meeting of the club, the research plan was perceived as being too unwieldy, complicated, and open-ended. Worst of all, the only interested source of funding had rejected it. By the end of July, the situation was again dramatically reversed. The Club of Rome had endorsed a new research plan, based on Forrester's ideas, and by mid-November the first phase of that plan, the *Limits to Growth* project, was fully funded.

The use of System Dynamics as the backbone of the *Limits to Growth* project was an arrangement mutually beneficial to Forrester and the Club of Rome. It allowed Forrester to extend his set of tools into wider realms of practice, and it allowed the Club of Rome to have a specific, well-delineated project that would eventually catapult them into the whirlwind of international discussions about global problems. The arrangement also involved very dramatic changes of orientation in the type of project the club sponsored: in the geographic location of the project, in its theoretical underpinnings, in the degree of financial support ultimately attracted, and in the very character of the Club of Rome itself.

Forrester transformed his role vis-à-vis the project: He moved from being a peripheral member of the club to being the key bearer of the theories and tools that were at the core of its first project. This transformation certainly involved historical contingency: Forrester was in the right place at the right time, just when the executive committee of the club needed a new way to move forward. It also involved Forrester's ability to turn this need to his advantage by first suggesting his methodology as an option and then by constructing an "obligatory point of passage" with his two-week seminar in Cambridge, during which time he demonstrated his system.[10] Forrester's demand that representatives of the Club of Rome come to Cambridge for two weeks "or not at all" allowed him to cement his ability to make his methodology the obligatory point of passage.[11] It was at that two-week seminar that he was able to convince his guests, Peccei being foremost, of the rightness and usefulness of his methods. He was able to concretely connect his methods with the history of research in servo-mechanisms at MIT, thus gaining the double support of his methodology having a long history and of its being tied to one of the foremost science universities in the world. Furthermore, he was able to have many of his colleagues give presentations on how System Dynamics was being applied to many areas of human endeavor with much apparent success. These presentations, in the aggregate, presented a picture of System Dynamics as being applicable to a wide range of subjects and problems.

The Cambridge meeting shifted the character of the Club of Rome's plan quite substantially, though neither Peccei nor other members of the club ever publicly recognized this fact. Besides changing the geographic venue of the first project from Geneva to Cambridge, Massachusetts, the meeting led to the abandonment of Ozbekhan's proposed plan to establish a small, ongoing research center in Europe. Instead, the Club of Rome undertook to sponsor a specific project, with a clear beginning and ending. In its history, up to the present, the Club of Rome has not tried to build that small research center that figured in the Ozbekhan proposal. In fact, in many ways the club has acted more as a general contractor, sponsoring projects by outside researchers that have specific target dates of completion. Forrester's methodology certainly carried the day with the Club of Rome. It was also successful in convincing a funding source, the Volkswagen Foundation, of its value and potential fruitfulness. Forrester's methodology as embodied in the Meadows proposal, coupled with the vigorous lobbying of foundation members by Pestel, produced the desired result, and funding was forthcoming.

How were the unsuccessful Ozbekhan proposal and the successful Meadows proposal different? Why was one able to gather support while the other was not? Certainly, the cogency of the Meadows presentation was significant. The proposal was long on specific methodology—buttressed by concrete examples of computer runs—and short on theoretical discussions about different possibilities. The whole proposal confidently assumed that its methodology was the right one for the job. Certainly, the ability of the Meadows team to do the job at less cost than what Ozbekhan called for must also have had an impact.

Beyond the cogency and self-assuredness of the Meadows presentation and beyond the issue of comparative costs, the two proposals embodied different basic assumptions. In order to explore this point further, a distinction made between "exogenous" and "endogenous" types of modeling feedback mechanisms is quite useful. In his book *Feedback Thought in Social Science and Systems Theory* (1991), George Richardson traces the origins of two distinct types of feedback thinking in the social sciences. One he calls the "cybernetic thread," and the other he calls the "servo-mechanism thread" (Richardson 1991, 93). Given that, according to Richardson, these modes of thinking are derived largely from engineering practices, he looks at the types of practices connected with each thread.[12] The "cybernetics thread" emerged from practices connected with communications, while the "servo-mechanisms thread" emerged from servomechanism control projects. The distinct problems associated with each set of practices produced different perceptions of the structure of a given system and how that system functioned. Richardson states:

> The two engineering fields differed in their perceptions of the sources of dynamics in the system. In a servomechanism, it is the *process of regulation* itself that shapes the dynamic behavior of the system in response to some arbitrary and largely uninteresting exogenous disturbance. In an amplifier, it is the *incoming wave form* that shapes the major characteristics of the output of the system. (Richardson 1991, 164)

If we keep in mind the basic differences that Richardson discusses in the above quotation regarding the sources of dynamics in a system when comparing the Ozbekhan proposal and the Meadows proposal, we can understand why one appeared more concrete than the other. For Ozbekhan, trained in the communications or cybernetics thread, the exogenous or external input is essential for an understanding of the system under consideration.[13] So, when studying global problems, he proposes a structure to gather information about the Continuous Critical Problems and to explore how they affect the system as a whole. He can conceive of modeling only after extensive understanding of the exogenous inputs to the world system, such as growing population or pollution. In the case of the Meadows (via Forrester) proposal, the exogenous causes are of no great interest. Rather, the focus is on the internal dynamics of the system. For Meadows, the necessity is to have a model of the system as soon as possible to focus on the internal dynamics. It is the study of those dynamics that will produce counterintuitive insights that allows planners to come to correct understandings. Thus, the Meadows proposal was highly focused and concrete, while the Ozbekhan proposal may have appeared as too wordy and vague.[14]

With funding secured, and with a clear methodology and goal at hand, the Club of Rome project on the predicament of mankind was finally under way.

The Meadows Team and the Limits to Growth

The global model that the Meadows team produced, known as "World3," was closely patterned on the "World2" model that Forrester had developed. In fact, according to Pestel, who was somehow disappointed by this fact, the Meadows model was quite similar to Forrester's. Pestel states:

> Meadows' world model, "World 3," which was in the making, hardly differed from Forrester's "World 2," the difference consisting chiefly in the larger number of auxiliary variables influencing the rate equations, which were backed by far more empirical evidence than in Forrester's work, and in the extensive use of delay functions, the latter being responsible for the steeped collapse modes in Meadows' model runs, as anyone acquainted with feedback control theory could readily expect. (Pestel 1989, 26)

Forrester published his *World Dynamics* while the Meadows team was still working on its *Limits to Growth* model. There was some apprehension among Club of Rome members that the Forrester book would somehow steal the public impact winds from the Meadows project's sails. These fears proved false. As Pestel dryly notes, "with the academic title *World Dynamics* (analogous to his former titles *Industrial Dynamics* and *Urban Dynamics*), Forrester's book gained only scant attention" (Pestel 1989, 27).[15] Pestel's comments point out the immense significance of what might be considered a mundane detail—the title of a book. Indifference to marketing was not a mistake that the Club of Rome was to make. Peccei, in particular, had a very clear understanding that marketing had a profound effect on the depth of impact of a particular action. The ultimate success of the *Limits to Growth* book was very dependent on its intelligent marketing. *Limits to Growth* was eventually to sell close to ten million copies worldwide, with translations in more than twenty languages.

While the marketing of the book was well planned and carried out very successfully, the very notion of the book *Limits to Growth* emerged quite accidentally. The original goal of the Meadows team was to produce a scientific report, along the lines of what eventually became *Dynamics of Growth in a Finite World* (Meadows et al. 1974). This book, published two years after *Limits to Growth*, included the scientific apparatus, such as the equations depicting the feedback relationships between the five fields composing the "World3" model, that was missing from *Limits to Growth*. In fact, it was the lack of a proper scientific apparatus that was used by many critics of *Limits to Growth*, particularly in the academic community, to disparage and minimize the book.[16]

The plan to orient the book toward lay readers can be traced to a general meeting of the Club of Rome in Montebello, Quebec, in April of 1971. The meeting was held at the invitation of Pierre Trudeau, then prime minister of Canada.[17] At that meeting Forrester did a presentation of some of the concepts of System Dynamics, and Meadows reported on his team's research progress.[18] During his presentation and afterward at more informal meetings with several members of the club, Meadows was surprised by how difficult it was for many of those people to understand some of the basic assumptions behind "World3." As Meadows stated it:

> I would use the term "exponential growth" and they would not understand what the term technically meant, or I would say "physical limits" and they were not willing to accept the notion that there are physical limits."[19]

At that point Meadows asked his wife Donella, the eventual writer of *Limits to Growth*, to come up with ten to twelve pages of definitions that could be handed out before these meetings to help people grasp the concepts behind System Dynamics. This initial paper was mailed to members of the club, and a "torrent of questions and objections and requests for further clarifications" came back. Donella Meadows revamped the original document to about 50 pages. The process was repeated, and more pages were added to a total of about 100 pages. In a three-page letter that Donella Meadows sent to Forrester, detailing the origin of *Limits*, she stated that after several revisions the document became too large for "the staple-in-the-left-hand-corner sort of document we had originally planned."[20] At around that time Peccei became insistent on the production of a general publication without having to wait for a technical report and was asking the Meadows to produce several thousand copies of their existing draft.[21] It was at this point, sometime in September of 1971, that the Meadows decided to work on the general book that was to become *Limits to Growth*.[22]

A decision was made to produce a book for the general public and only later to publish a more technical report, and a publisher was sought. Dennis Meadows sent a copy of one of the versions of the manuscript to two groups that handled policy-oriented publications. These groups were the Overseas Development Council (ODC) and Potomac Associates.[23] Potomac Associates took an immediate interest in the project and offered to handle all aspects of bringing the book to the public.[24] Increasingly overwhelmed by having to publish the book on top of their ongoing research, the Meadows decided to accept the Potomac Associates offer.

The book was formally published and distributed by Universe Books, a publishing house used by Potomac Associates. The book came out in March of 1972, later than Peccei had hoped, but much earlier than would have been the case for a scientific monograph.[25] The *Limits to Growth* project was to last roughly twenty months, from July 1970 to the first week in March of 1972. After *Limits to Growth* came out, the Club of Rome in general and Peccei in particular quickly lost interest in the more technical part of the project. They had in their hands the kind of publication they wanted, with even more worldwide impact than they ever dreamed. According to Meadows, after March of 1972, the club lost interest in the proposed technical report to be released. From that point onward the club was basically interested in the Meadows as speakers rather than as scientific researchers.[26]

Soon after *Limits to Growth* appeared, the research team under Meadows was formally disbanded, though most members became involved in further modeling work (Meadows et al. 1974). In the next couple of years, the Meadows did complete the more technical report of the original research and produced two other related publications. The technical report came out as *Dynamics of Growth in a Finite World* (Meadows et al. 1974). It included a technical description of "World3," plus the empirical data. The other two books were *Toward Global Equilibrium: Collected Papers*, published in 1973, including individual papers by members of the System Dynamics group at MIT, and *Alternatives to Growth*, published in 1977, containing the winning papers from a competition on sustainable futures.

Limits to Growth and the New Governmentality

The launching of *Limits to Growth* was a masterful exercise in marketing. It quickly generated interest that snowballed to global proportions.[27] It launched a new field of scientific practice, namely, global computer modeling, and focused attention, as nothing had done before, on the close relationship between human production and consumption and the planet as a whole.

In the process of doing the computer modeling work, the Meadows team sought feedback on their ongoing research. Eventually, word about their investigations reached a wider audience. At one point, for example, a Dutch journalist produced a television program on some of the research prior to *Limits to Growth*'s formal publication date (Moll 1991, 99). By and large, however, the wide impact of the work did not appear until Potomac Associates formally launched the book in March of 1972. They did so by organizing a conference in Washington, DC, in conjunction with the Club of Rome, the Smithsonian Institution, and the Woodrow Wilson Center for Scholars. The costs of setting up the conference were covered largely by the Xerox Corporation.[28]

The target audience for the conference, consonant with the top-down approach favored by the Club of Rome, was composed of high-ranking government officials as well as members of academia and industry. About 250 individuals heard a presentation by Dennis Meadows, Peccei, King, and Pestel, as well as one by the then Secretary of Health, Education and Welfare, Elliot L. Richardson. The speakers were followed by a panel discussion which included Philip Abelson, then the editor of *Science*, and Lester Brown, senior fellow at the ODC, among others. The conference concluded with remarks by Carroll Wilson.[29]

In addition to the conference, Potomac Associates and the Club of Rome sent out over 15,000 copies of *Limits to Growth* to influential members of government, industry, and academia. The English version of the report was followed in quick succession with versions in German, French, Dutch, Italian, and Spanish.[30] In terms of marketing, the strategies were a complete success. Soon the club was spending most of its time engaged in worldwide debates about the report.

The book form of the research done by the Meadows team was published, in nontechnical form, as *Limits to Growth: A Report for the Club of Rome's Project on the Predicament of Mankind* (Meadows et al. 1972). The size of the book was relatively small, with slightly over 200 pages, including forty-eight illustrations and six numerical tables.[31] The bulk of the book consisted of an account of the Meadows team's research, written largely by Donella Meadows.[32] It also included a twelve-page commentary from the Club of Rome itself, signed by the six members of the executive committee (Meadows et al. 1972, 185). The commentary called for a heeding of the warnings embedded in the *Limits to Growth* report and the creation of a world forum "where statesmen, policy-makers, and scientists can discuss the dangers and hopes for the future global system without the constraints of formal intergovernmental negotiation" (Meadows et al. 1972, 197).

The *Limits to Growth* report is itself divided into an introduction and five chapters. The introduction contains general discussion of how the space and timescale of problems confronting humanity overwhelms the local and short-time perspective within which most human beings operate. Even within a narrow time and space perspective, the authors argue, human beings use models to plan ahead. A farmer, for example, uses a mental model of past and present weather conditions, market trends, and the land he or she holds to plan future crops. These mental models tend to be simple and relatively short-term and are quickly overwhelmed by complexity (Meadows et al. 1972, 21). The *Limits to Growth* project, by contrast, contains a formal, written model of the interactions of five major trends, global in nature, operating over long periods of time. It has the advantages, over everyday mental models, of being amenable to replication of its results and is able to accommodate issues of complexity on a much larger scale. The introduction generally argues that problems are now global in nature and are too complex to be understood through the usual mental modeling that humans use. On the other hand, the introduction asserts, the use of formal models does not represent a radical departure from what humans have done all alone. Given the rather stark conclusions of *Limits to Growth,* it is understandable that the introduction minimizes the discontinuity of practice that global modeling represents. However, it is important to keep in mind the novel character of the methodology of global computer modeling that *Limits to Growth* advances.[33]

The introduction to the book also includes a list of provocative conclusions, which read in part:

> If the present growth trends in world population, industrialization, pollution, food production, and resource depletion continue unchallenged, the limits to growth on this planet will be reached sometime within the next one hundred years. The most probable result will be a rather sudden and uncontrollable decline in both population and industrial capacity. (Meadows et al. 1972, 24)

This stark prediction quoted above sets the stage for the book as a whole. Thematically, the book can be broadly divided into three parts. The first, corresponding to chapters 1 and 2, discusses the differences between linear and exponential growth in general, with specific focus on population and industrial growth. The authors highlight the suddenness of change under a regime of exponential growth by describing a French riddle for children: A water lily is growing in a pond by doubling in size everyday for thirty days. You decide that you will not worry about the growth covering the whole pond until it reaches half of the total surface. What day will that be? The answer is the twenty-ninth day. In other words, you only have one day in which to prevent the total choking of the lake by the growth of the water lily (Meadows et al. 1972, 29).

The authors argue that these processes of exponential growth have at their core a positive feedback loop. A "vicious circle" is also an example of a positive feedback loop. When looking at the growth of population and industrial output, we see that they are dominated by some type of specific positive feedback loop and negative feedback loop. In the case of population growth, improvements in sanitation and nutrition have led to a marked decrease in death rates (which would normally operate as a negative feedback loop), and this, in turn, has accelerated growth. In the case of industrial production, increasing demand for consumer goods has led to the increased production of capital goods (i.e., machines capable of producing more consumer goods), another type of vicious circle. The negative feedback loop, the wearing out of capital goods, is much smaller than the production of replacement capital goods. There is thus an exponential growth in industrial output (Meadows et al. 1972, 38).

In the first part of the report, the Meadows team also considers the constitutive positive and negative feedback loops of three other trends: food, nonrenewable resources, and pollution. Where food is concerned they conclude that the limits of food production needed to sustain the growing population of the planet will be reached by the year 2000. With respect to nonrenewable resources, they conclude that "given present resource consumption rates and the projected increase in these rates, the great majority of the currently important nonrenewable resources will be extremely costly 100 years from now" (Meadows et al. 1972, 66). Finally, with regard to pollution the authors do not come to definite conclusions about the ultimate limits of Earth's capacity to absorb it, but they warn of reaching these limits unexpectedly and dramatically.

The second part of the book, comprising chapters 3 and 4, is the heart of the computer modeling research. It is here that the authors explore, through their modeling, the interrelationships of the five trends that they had first discussed individually. In order to establish the causal relationships and feedback loop structures between the five trends, they consulted experts in the respective areas, as well as relevant literature in fields such as ecology, demography, and nutrition. Out of this research they established what they thought were the more important relations and quantified them by using available data from the World Bank, the United Nations, and related institutions (Meadows et al. 1972, 90).[34] The relations and quantifications were then translated into the DYNAMO equations that ultimately comprised the *Limits to Growth* model.[35] With the model in place, they produced several runs based on the assumptions of several existing policies about such issues as food production, pollution controls, and birth control (Meadows et al. 1972, 91).

The most dramatic run is the "standard" run. It assumes no major changes in the way humans have been relating with the environment for the last 100 years. The behavior of the model under those conditions is one of overshoot and collapse. The catastrophic ending is caused by depletion of nonrenewable resources. As the authors put it:

> The industrial capital stock grows to a level that requires an enormous input of resources. In the very process of that growth it depletes a large fraction of the resource reserves available. As resource prices rise and mines are depleted, more and more capital must be used for obtaining resources, leaving less to be invested for future growth. Finally investment cannot keep up with the depreciation, and the industrial basis collapses, taking with it the service and agricultural systems, which have become dependent on industrial inputs (such as fertilizers, pesticides, hospital laboratories, computers, and especially energy for mechanization). . . . Population finally decreases when the death rate is driven upward by lack of food and health services. (Meadows et al. 1972, 125)

In addition to the "standard" run, the authors tried out other alternatives, such as doubling or even quadrupling the known availability of resources or application of new technologies. The essential conclusion for each run remained the same: eventually the global system would overshoot and collapse. The reason for this was to be found, according to the Meadows team, in the complex feedback loops that existed among the different trends. Improvement in one or two trends would produce unexpected positive feedback loops in another trend, ultimately leading to systemic collapse. Thus, in one example they modeled an assumed reduction in the level of global pollution by a level of four. This is how they described the result of that run:

> The pollution control policy is indeed successful in reducing the pollution crisis of the previous run. Both population and industrial output per person rise well beyond their peak values . . . and yet resource depletion and pollution never become serious problems. The overshoot mode is still operative, however, and the collapse comes about this time primarily from food shortage. (Meadows et al. 1972, 137)

In the third and final part of the report, published as the last chapter of the book, the authors change direction to explore some of the changes that would be necessary to achieve global equilibrium and sustainability. They search for a global system that would be capable of satisfying the basic material needs of humans and would be sustainable in the long run, without a tendency toward collapse (Meadows et al. 1972, 158). They argue that of the three possibilities that most people think are possible, that is, continuous growth, self-imposed growth limits, and nature-imposed limits, only the last two are realistic. They conclude that, through self-imposed limits on growth, a transition from growth to global equilibrium and sustainability is possible, in fact, mandatory, if long-term survival of the global system is desired. The Meadows team concluded their report by stating that without a long-term goal and commitment toward global sustainability, "short-term concerns will generate the exponential growth that drives the world system toward the limits of the earth and ultimate collapse. With that goal and that commitment, mankind would be ready now to begin a controlled, orderly transition from growth to global equilibrium" (Meadows et al. 1972, 184).

Beyond the Limits

In 1992, three members of the original *Limits to Growth* research team plus three newcomers collaborated on a sequel project, entitled *Beyond the Limits.* The conclusions to which they arrived were quite similar to those embodied in the original report. To some extent the conclusions were somewhat more dire. They argued that

> the human use of many essential resources and generation of many kinds of pollutants have already surpassed rates that are physically sustainable. Without significant reductions in material and energy flows, there will be in the coming decades an uncontrolled decline in per capita food output, energy use, and industrial production. (Meadows, Meadows, and Randers 1992, xv)

In a departure from the original report, *Beyond the Limits* calls for more specific "systemic structural changes." By these terms they mean

> changing the *information* links in a system: the content and timeliness of the data that actors in the system have to work with, and the goals, incentives, costs, and feedbacks that motivate or constrain behavior. The

same combination of people, institutions, and physical structures can behave completely differently, if its actors can see a good reason for doing so and if they have the freedom to change. In time a system with new information structure can socially and physically transform itself. It can develop new institutions, new rules, new buildings, people trained for new functions. That transformation can be natural, evolutionary, and peaceful. (Meadows, Meadows, and Randers 1992, 191)

The shift from the original *Limits to Growth*, with its concern for the ultimate quantifiable physical limits of the planet's resources, to the newer *Beyond the Limits*, with its emphasis on the possibility of changing the structural nature of the global system, may signal to us the real contribution of the original work. This contribution lies in the fact that *Limits to Growth* opened the public imaginary to the possibility of thinking anew about the relation between humanity and the biosphere, a new thinking that dared to consider that relation as global and embodied in an irreducible system. The last statement from *Beyond the Limits* quoted above assumes a familiarity with the notion of a global system that simply did not exist twenty years ago. And this very fact illustrates the most important accomplishment of the *Limits to Growth* report, namely, advancing the idea that we live in the age of globality, the age of the global earth.

Notes

1. Forrester, interview by the author, December 12, 1991.

2. The information on the fee can be found in a budget submitted to the Volkswagen Foundation in November 1970. See the preliminary draft dated August 13, 1970, in the Gordon Stanley Brown Papers, MC24-17-686.

3. Meadows, interview by the author, May 31, 1994.

4. Ibid. A list of the seventeen members of the team and their areas of expertise can be found on page 6 of *Limits to Growth* (Meadows et al. 1972).

5. The general youthfulness of the research group can be assessed from the age of the team's leader. Dennis Meadows was 28 years old when he started the project.

6. I discuss Forrester's claim to universality in chapter 3.

7. I discuss these elements in more detail in chapter 3.

8. The budget is not included in the copy I have been using here, because it is a "public" version. This version is similar to the "member of CoR [Club of Rome]" versions except that it does not have a budget. The budget is taken from the August 13 draft. The "public" version, dated November 6, 1970, can be found in the Carroll Louis Wilson Papers, MC29-21-935.

9. The two German scientists were Erich Zahn and Peter Milling. A copy of the Pestel letter is located in the Gordon Stanley Brown Papers, MC24-17-683.

10. I borrow this term, with more or less the same meaning, from Latour (1988). He argues, in discussing Pasteur, that Pasteur was able to turn his laboratory into a point of passage, through which politicians, hygienists, farmers, and microbes had to go in order for certain effects, such as better health, to occur.

11. Forrester, interview by the author, December 12, 1991.

12. I follow a path similar to Richardson's in exploring the emergence of System Dynamics out of the practices of World War II (Richardson 1991). Richardson has extensive discussions of System Dynamics, and I have benefited greatly from his book. Richardson is a practitioner of System Dynamics and studied with Forrester at MIT.

13. There is no doubt that Ozbekhan's work is largely influenced by the cybernetic thread. This was the theoretical focus at SDC where he was director of planning.

14. Ozbekhan never came to terms with Forrester's approach. He was a member of the Club of Rome delegation that went to Cambridge to study the

System Dynamics approach. According to Moll, he disappeared from the scene after three days. As Moll describes it from an interview with Thiemann: "As the majority of the Executive Committee was positive, Peccei turned to Ozbekhan and told him that most people present thought that it would be advisable to use Forrester's methodology, and, as he (Peccei) himself thought very highly about his conceptual work, he wondered whether Ozbekhan would be prepared to cooperate with Forrester in leading the project. Ozbekhan, who had been sounding since the Bern meeting as if he would eventually be prepared to cooperate . . . asked for a couple of hours to think about it. Ten minutes later though he returned to the room, said that this was all a lot of crap and quit the whole undertaking together with his membership in the Club" (Moll 1991, 79).

15. In an autobiographical essay Forrester paints a more positive picture of the reception of *World Dynamics*. He writes that "the book seemed to have everything necessary to guarantee no public notice. First, it had forty pages of equations in the middle that should be sufficient to squelch public interest. Second, the main messages were presented as computer output graphs, and most of the public does not understand such presentations. Third, the publisher of the book had published only one previous book and I doubted that *World Dynamics* had the commercial status even to be reviewed. I intended the book for maybe 20 people in the world who would like to study an interesting model in their computers. . . . *World Dynamics* came out the first week of June, 1971. During the last week of June, it was reviewed in the *London Observer,* which then circulated around the world. A letter from a professor in New York asked for more information because he had been reading about the book in the *Singapore Times!* In August the book had the full front page of the second section of the *Christian Science Monitor,* in September a page and a half in *Fortune,* and in October a column in the *Wall Street Journal*" (Forrester 1991, 22). The fact remains that, as Pestel tells it, the book did not have a huge impact, and it would be completely eclipsed by *Limits to Growth.* This fact Forrester ruefully concedes, when referring to a very particular article: "*World Dynamics* was the subject of a full-length article in *Playboy.* But, as a communications medium for conveying System Dynamics, that magazine was a disappointment. Out of eight million copies printed, the only response I received was a request to conduct a two-day meeting for the Board of Overseas Missions after the article was read by a man at the National Council of Churches" (Forrester 1991, 22). The book would receive a withering critique in 1973 by William D. Nordhaus in a long review in the prestigious *Economic Journal* of the Royal Economic Society (Nordhaus 1973). For Forrester's response, see "The Debate on *World Dynamics:* A Response to Nordhaus" (1974).

16. See, for example, the article by Robert Gillette (1972).

17. A copy of the formal convocation of the meeting can be found in the Carroll Louis Wilson Papers, MC29-54-2093.

18. A copy of the postmeeting report prepared by Peccei can be found in the Carroll Louis Wilson Papers, MC29-38-1574.

19. Meadows, interview by the author, May 31, 1994.

20. A copy of the letter can be found in the Gordon Stanley Brown Papers,

MC24-17-683. The letter was forwarded, with a cover letter dated May 7, 1972, by Forrester to Jerome Wiesner, then president of MIT. Apparently Wiesner had started to get an unusual number of letters about *Limits to Growth*. The Forrester and Meadows letters were intended to get Wiesner up to speed about the subject, since MIT figured prominently in the growing debate over the conclusions of the Club of Rome report.

21. Donella Meadows, letter to Forrester, May 4, 1972, in the Gordon Stanley Brown Papers, MC24-17-683.

22. Meadows, interview by the author, May 31, 1994.

23. At ODC Dennis Meadows was familiar with Lester Brown, who would later found the World-Watch Institute.

24. The information on the process of finding a publisher comes from Donella Meadows's letter to Forrester and from my interview with Dennis Meadows. Potomac Associates is a for-profit publishing venture started in August of 1970 with seed money from Irwin Miller, then chairman of the board of Cummins Engine. It was run by William Watts and Donald Lesh, both former Foreign Service officers and friends of the Meadows. The goal of Potomac Associates has been to publish public policy–oriented studies. *Limits to Growth* was their second project. The first was a study entitled *Hopes and Fears of the American People*. Information on Potomac Associates comes from a five-page document by Lesh detailing the history of their involvement with *Limits to Growth*. A copy can be found in the Gordon Stanley Brown Papers, MC24-17-683.

25. In a letter to the executive committee of the club dated October 27, 1971, Peccei expresses some frustration at the ever-moving date of completion of the *Limits to Growth* project. He writes: "Dennis does not appreciate, as for instance I do, the necessity of not delaying the publication of this Report—which should have been ready in July or August, but which will be practically out only in January 1972." As it happened, the book came out even later. For Peccei's letter, see the Carroll Louis Wilson Papers, MC29-54-2093.

26. Meadows, interview by the author, May 31, 1994.

27. I will not rehearse the debate over the report since it has been amply covered elsewhere. See, for example, Sandbach (1978a, 1978b), McCutcheon (1979), and Moll (1991).

28. See "Potomac Associates and the Limits to Growth" by Lesh for organization details of the conference and other marketing strategies, in the Gordon Stanley Brown Papers, MC24-17-683.

29. A copy of the proceedings of the conference, about 145 pages, can be found in the Carroll Louis Wilson Papers, MC29-55-2101.

30. Details of the coordinated launching of the several translations can be found in a letter from Peccei, dated December 31, 1971, to all members of the Club of Rome, in the Carroll Louis Wilson Papers, MC29-54-2095. The copyrights for all languages, except English, were owned by the club. The English copyright was owned by Dennis Meadows and Potomac Associates. The translated versions were not without some problems. The French one, for example, was handled by one Janine Delauney who, according to Dennis Meadows, added about fifty pages

of her own commentary to the book (Meadows, interview by the author, May 31, 1994).

31. By comparison, the technical report of the project, entitled *Dynamics of Growth in a Finite World* (Meadows et al. 1974), topped 600 pages.

32. Meadows, interview by the author, May 31, 1994.

33. *Limits to Growth* represents the beginning of the practice of global computer modeling. Many projects were launched after this initial effort by the Meadows group. This type of modeling is still widely practiced, but its character has changed, with climatic effects of anthropogenic change being the newer focus, at the expense of attempting to model a wider number of variables, such as capital accumulation and malnutrition. In other words, modeling has become less social and more biological and climatological. This trend, of course, dovetails nicely with my claim that we are moving from a form of governmentality preoccupied with "man" (the social) to a governmentality focused on the biosphere itself. A discussion of global modeling since the original effort by the Meadows team can be found in "Global Models: A Review of Recent Developments" by Sam Cole (1987).

34. The information on sources of data came from my interview with Dennis Meadows, on May 31, 1994.

35. DYNAMO is a programming language used for simulations. Details about the assumptions behind the equations as well as more information about the empirical data used can be found in *Dynamics of Growth in a Finite World* (Meadows et al. 1974).

Bibliography

Books and Articles

Ahmad, Yusuf, Salah E. Serafy, and Ernst Lutz, eds. 1989. *Environmental Accounting and Sustainable Development: A UNEP-World Bank Symposium.* Washington, DC: World Bank.

Albrow, Martin. 1996. *The Global Age: State and Society beyond Modernity.* Oxford: Polity.

Baker, James D. 1990. *Planet Earth: The View from Space.* Cambridge, MA: Harvard University Press.

Barney, Gerald O., ed. 1981. *The Global 2000 Report to the President of the United States.* 2 vols. Elmsford, NY: Pergamon.

Baudrillard, Jean. 1986. "The Year 2000 Will Not Happen." In *Futur*Fall: Excursions into Post-Modernity,* edited by E. A. Grosz, T. Threadgold, D. Kelly, A. Cholodenko, and E. Colless. Sidney: Power Institute of Fine Arts.

Baum, Claude. 1981. *The System Builders: The Story of SDC.* Santa Monica, CA: System Development Corporation.

Baxter, James Phinney. 1946. *Scientists against Time.* Cambridge, MA: MIT Press.

Beck, Ulrich. 1992. "From Industrial Society to Risk Society: Questions of Survival, Social Structure and Ecological Enlightenment." *Theory, Culture and Society* 9:97–123.

Bergman, Charles. 1990. *Wild Echoes.* New York: McGraw-Hill.

Bloomfield, Brian. 1986. *Modelling the World: The Social Constructions of Systems Analysts.* New York: Basil Blackwell.

Bowker, Geoff. 1993. "How To Be Universal: Some Cybernetic Strategies, 1943–70." *Social Studies of Science* 23:107–23.

Brabyn, Howard. 1972. "Cool Catalyst." *New Scientist* 55 (August 24, 1972): 390–91.

Brown, Gordon, and Donald P. Campbell. 1948. *Principles of Servomechanisms.* New York: John Wiley and Sons.

Burchard, John. 1948. *MIT in World War II.* New York: John Wiley and Sons.

Burchell, Graham, Colin Gordon, and Peter Miller, eds. 1991. *The Foucault Effect: Studies in Governmentality.* Chicago: University of Chicago Press.

Buttel, Frederick H., Ann P. Hawkins, and Alison G. Power. 1990. "From Limits to Growth to Global Change." *Global Environmental Change* 1 (December): 57–66.

Clark, William C. 1986. "Sustainable Development of the Biosphere: Themes for a Research Program." In *Sustainable Development of the Biosphere*, edited by William C. Clark and R. E. Munn, 5–47. New York: Cambridge University Press on behalf of the IIASA.

Clark, William C., and R. E. Munn, eds. 1986. *Sustainable Development of the Biosphere*. New York: Cambridge University Press on behalf of the IIASA.

Cohen, I. Bernard. 1988. "The Computer: A Case Study of Support by Government, Especially the Military, of a New Science and Technology." In *Science, Technology and the Military*, edited by E. Mendelsohn, M. R. Smith, and P. Weingart, 119–54. Sociology of the Sciences vol. 12. Boston: Kluwer Academic.

Cole, Sam. 1987. "Global Models: A Review of Recent Developments." *Futures* 19 (August): 403–30.

Crary, Jonathan. 1990. *Techniques of the Observer*. Cambridge, MA: MIT Press.

de Lemos, Harold Mattos. 1990. "Amazonia: In Defense of Brazil's Sovereignty." *Fletcher Forum of World Affairs* 14, no. 2:301–12.

Deleuze, Gilles. 1992. "Postcript on the Societies of Control." *October* 59 (Winter): 3–7.

Edwards, Paul N. 1996. *The Closed World: Computers and the Politics of Discourse in Cold War America*. Cambridge, MA: MIT Press.

Elichirigoity, F. I. 1992. "Towards a Genealogy of Planet Management: The Role of Scientific Practices." Paper presented at the workshop, The Components of Practice, November 6, 1992, London School of Economics.

———. 1994. "Towards a Genealogy of Planet Management: Computer Simulation, Limits to Growth and the Emergence of Global Spaces." Ph.D. diss., University of Illinois at Urbana-Champaign.

Escobar, Arturo. 1987. "Power and Visibility: Development and the Invention and Management of the Third World." *Cultural Anthropology* 3, no. 4:428–43.

Everett, Robert R., ed. 1983. Special issue: "Sage." *Annals of the History of Computing* 5, no. 4 (October 1983).

Ferguson, James. 1990. *The Anti-Politics Machine: "Development," "Depoliticization, and Bureaucratic Power in Lesotho*. New York: Cambridge University Press.

Forrester, Jay W. 1958. "Industrial Dynamic: A Major Breakthrough for Decision Makers." *Harvard Business Review* 36, no. 4:37–66.

———. 1961. *Industrial Dynamics*. Cambridge, MA: Productivity.

———. 1965. "A New Corporate Design" *Sloan Management Review* 7, no. 1:5–17.

———. 1969a. "A New Corporate Design." In *Perspectives of Planning*, edited by Erich Jantsch, 425–48. Paris: OECD.

———. 1969b. "Planning Under the Dynamic Influences of Complex Social Systems." In *Perspectives of Planning*, edited by Erich Jantsch, 237–56. Paris: OECD.

———. 1969c. *Urban Dynamics*. Cambridge, MA: Productivity.

———. 1973a. "Counterintuitive Behavior of Social Systems." In *Toward Global Equilibrium: Collected Papers*, edited by Dennis Meadows and Donella Meadows, 3–30. Cambridge, MA: Wright-Allen.

———. 1973b. *World Dynamics*. 2d ed. Cambridge, MA: Productivity.

———. 1974. "The Debate on *World Dynamics:* A Response to Nordhaus." *Policy Sciences* 5:169–90.

———. 1975. *Collected Papers of Jay Forrester*. Cambridge, MA: Wright-Allen.

———. 1991. "From the Ranch to System Dynamics: An Autobiography." In *Management Laureates: A Collection of Autobiographical Essays*, edited by Arthur Bedeian. New York: JAI.

Fortun, Mike, and S. S. Schweber. 1993. "Scientists and the Legacy of World War II: The Case of Operations Research (OR)." *Social Studies of Science* 23:595–642.

Foucault, Michel. 1970. *The Order of Things*. New York: Vintage.

———. 1991a. "On Governmentality." In *The Foucault Effect*, edited by Graham Burchell, Colin Gordon, and Peter Miller, 87–104. Chicago: University of Chicago Press.

———. 1991b. "Questions of Method." In *The Foucault Effect*, edited by Graham Burchell, Colin Gordon, and Peter Miller, 75–86. Chicago: University of Chicago Press.

Frosch, Robert A., and Nicholas E. Gallopoulos. 1989. "Strategies for Manufacturing." *Scientific American* 261, no. 3 (September): 144–52.

Geyer, Michael, and Charles Bright. 1995. "World History in a Global Age." *American Historical Review* 100 (October): 1034–60.

Ghilarov, Alexej M. 1995. "Vernadsky's Biosphere Concept: An Historical Perspective." *Quarterly Review of Biology* 70, no. 2:193–203.

Gill, Stephen. 1990. *American Hegemony and the Trilateral Commission*. New York: Cambridge University Press.

Gillette, Robert. 1972. "The Limits to Growth: Hard Sell for a Computer View of Doomsday." *Science* 175 (March): 1088–90.

Gleick, Peter. 1989. "The Implications of Global Climatic Change for International Security." *Climatic Change* 15, no. 1/2:309–25.

Golub, Robert, and Joe Townsend. 1977. "Malthus, Multinationals and the Club of Rome." *Social Studies of Science* 7:201–22.

Grunwald, Joseph, and Kenneth Flamm. 1985. *The Global Factory*. Washington, DC: Brookings Institute.

Gwynne, Michael, and Wayne Moneyhan. 1989. "The Global Environment Monitoring System and the Need for a Global Resource Database." In *Changing the Global Environment: Perspectives on Human Involvement*, edited by Daniel Botkin, Margriet Caswell, John Estes, and Angelo Orio. Boston: Academic.

Haraway, Donna. 1981–82. "The High Cost of Information in Post–World War II Evolutionary Biology: Ergonomics, Semiotics, and the Sociobiology of Communications Systems." *Philosophical Forum* 13, nos. 2–3:244–78.

———. 1985. "A Manifesto for Cyborgs." *Socialist Review* 15, no. 2:65–108.

———. 1989. *Primate Visions: Gender, Race, and Nature in the World of Modern Science*. New York: Routledge.

Hecht, Susanna B., and Alexander Cockburn. 1990. *The Fate of the Forest: Developers, Destroyers, and Defenders of the Amazon*. New York: HarperPerennial.

Heims, J. Steve. 1991. *The Cybernetics Group*. Cambridge, MA: MIT Press.

Hershberg, James. 1993. *James B. Conant: Harvard to Hiroshima and the Making of the Nuclear Age.* New York: Knopf.

Hewlett, G. Richard, and Francis Duncan. 1972. *A History of the United States Atomic Energy Commission.* Vol. 2. Washington, DC: Atomic Energy Commission.

Homer-Dixon, Thomas. 1991. "On the Threshold: Environmental Changes as Causes of Acute Conflict." *International Security* 16, no. 2:76–116.

Jantsch, Erich. 1967. *Technological Forecasting in Perspective.* Paris: OECD.

———, ed. 1969. *Perspectives of Planning.* Paris: OECD.

Kaplan, Fred M. 1991. *The Wizards of Armageddon.* New York: Simon and Schuster, 1983. Reprint, with a new foreword, Stanford, CA: Stanford University Press.

Keller, Evelyn Fox. 1995. *Refiguring Life: Metaphors of Twentieth-Century Biology.* New York: Columbia University Press.

Kelley, Kevin W. 1985. *The Home Planet.* Reading, MA: Addison-Wesley.

Kelly, Kevin. 1994. *Out of Control: The Rise of Neo-Biological Civilization.* Reading, MA: Addison-Wesley.

Keyfitz, Nathan. 1989. "The Growing Human Population." *Scientific American* 261, no. 3:118–27.

Killian, James R. 1985. *The Education of a College President: A Memoir.* Cambridge, MA: MIT Press.

King, Alexander. 1981. "The Club of Rome and Its Policy Impact." *Knowledge and Power in Global Society,* edited by William M. Evan, 205–24. Beverly Hills, CA: Sage.

Kwa, Chunglin. 1987. "Representations of Nature Mediating between Ecology and Science Policy: The Case of the International Biological Programme." *Social Studies of Science* 17:413–42.

Latour, Bruno. 1987. *Science in Action.* Cambridge, MA: Harvard University Press.

———. 1988. *The Pasteurization of France.* Cambridge, MA: Harvard University Press.

———. 1993. *We Have Never Been Modern.* Translated by Catherine Porter. Cambridge, MA: Harvard University Press.

Levitt, Theodore. 1988. "The Pluralization of Consumption." *Harvard Business Review* 66 (May–June): 7–8.

Lipschutz, Ronnie D. 1992. "Reconstructing World Politics: The Emergence of Global Civil Society." *Millennium: Journal of International Studies* 21, no. 3:389–420.

Lovelock, James. 1979. *Gaia: A New Look at Life on Earth.* New York: Oxford University Press.

———. 1986. "Gaia: The World as Living Organism." *New Scientist* (December 18): 25–28.

———. 1988. *The Ages of Gaia.* New York: Norton.

Lyman, Francesca. 1989. "What Gaia Has Brought." *Technology Review* 92, no. 5 (July): 53–61.

Lyotard, Jean-François. 1984. *The Postmodern Condition: A Report on Knowledge.* Minneapolis: University of Minnesota Press.

MacNeill, Jim, Peter Winsemius, and Taizo Yakushiji. 1991. *Beyond Interdependence:*

The Meshing of the World's Economy and the Earth's Ecology. New York: Oxford University Press.

Magretta, Joan. 1997. "Growth through Global Sustainability: An Interview with Monsanto's CEO, Robert B. Shapiro." *Harvard Business Review* 75 (January–February): 78–88.

Mathews, Jessica. 1989. "Redefining Security." *Foreign Affairs* 68, no. 2 (Spring): 162–77.

Mazlish, Bruce. 1993. *The Fourth Discontinuity: The Co-Evolution of Humans and Machines.* New Haven, CT: Yale University Press.

McCutcheon, Robert. 1979. *Limits of a Modern World: A Study of the "Limits to Growth" Debate.* London: Butterworths.

Meadows, Dennis. 1970–71. "Project on the Predicament of Mankind—Research Proposal." Club of Rome.

————, ed. 1977. *Alternatives to Growth.* Cambridge, MA: Ballinger.

Meadows, Dennis, William Behrens, Donella Meadows, Roger Naill, Jorgen Randers, and Erich Zahn. 1974. *Dynamics of Growth in a Finite World.* Cambridge, MA: Wright-Allen.

Meadows, Dennis, and Donella Meadows. 1973. *Toward Global Equilibrium: Collected Papers.* Cambridge, MA: Wright-Allen.

Meadows, Donella, Dennis Meadows, Jorgen Randers, and William Behrens III. 1972. *The Limits to Growth: A Report for the Club of Rome's Project on the Predicament of Mankind.* New York: Universe.

Meadows, Donella, Dennis L. Meadows, and Jorgen Randers. 1992. *Beyond the Limits.* Post Mills, VT: Chelsea Green.

Mendes, Candido. 1992. "The Quest for Globality: Interdisciplinary Endeavor in the Social and Natural Sciences." *International Social Science Journal* 132: 602–14.

Mesarovic, Mihajlo, and Eduard Pestel. 1974. *Mankind at the Turning Point.* New York: E. P. Dutton.

Miller, Peter, and Ted O'Leary. 1987. "Accounting and the Construction of the Governable Person." *Accounting, Organization and Society* 12, no. 3:235–67.

Miller, Peter, and Nikolas Rose. 1990. "Governing Economic Life." *Economy and Society* 19, no. 1:1–31.

Moll, Peter. 1991. *From Scarcity to Sustainability, Future Studies and the Environment: The Role of the Club of Rome.* New York: Peter Lang.

Morison, Samuel Eliot. 1963. *The Two-Ocean War.* Boston: Atlantic Monthly.

Morse, Philip M. 1977. *In at the Beginnings: A Physicist's Life.* Cambridge, MA: MIT Press.

Morse, Philip M., and George Kimball. 1951. *Methods of Operations Research.* London: Chapman and Hill.

Mounsey, Helen, and Roger F. Tomlinson, eds. 1988. *Building Databases for Global Science.* London: Taylor and Francis.

Myers, Norman. 1989. "Environment and Security." *Foreign Policy* 74 (Spring): 23–41.

Noble, David F. 1984. *Forces of Production: A Social History of Industrial Automation.* New York: Oxford University Press.

Nordhaus, William D. 1973. "World Dynamics: Measurement without Data." *Economic Journal* 83 (December): 1156–83.

Ohmae, Kenichi. 1989. "Managing in a Borderless World." *Harvard Business Review* 67 (May–June): 152–61.

Ozbekhan, Hasan. 1969. "Toward a General Theory of Planning." In *Perspectives of Planning*, edited by Erich Jantsch, 47–55. Paris: OECD.

———. 1970. "Quest for Structural Responses to Growing World-wide Complexities and Uncertainties." Club of Rome.

———. 1976. "The Predicament of Mankind." In *World Modeling: A Dialogue*, edited by C. West Churchman and Richard O. Mason, 11–25. New York: Elsevier.

Pauli, Gunter A. 1987. *Crusader for the Future: A Portrait of Aurelio Peccei, Founder of the Club of Rome.* New York: Pergamon.

Peccei, Aurelio. [1965] 1987. "The Challenge of the 1970's for the World of Today." Paper presented at the National Military College, Buenos Aires, September 27. Reprinted in *Crusader for the Future* by Gunter Pauli. New York: Pergamon.

———. 1969a. *The Chasm Ahead.* New York: Macmillan.

———. 1969b. "Reflections on Bellagio." In *Perspectives of Planning*, edited by Erich Jantsch, 517–19. Paris: OECD.

———. 1977. *The Human Quality.* New York: Pergamon.

Pestel, Eduard. 1989. *Beyond the Limits to Growth.* New York: Universe.

Pickering, Andrew. 1992. "The New Rationality." Manuscript, Cambridge University.

———. 1995. "Cyborg History and the World War II Regime." *Perspectives in Science* 3, no. 1:1–48.

Pickering, Andrew, and Adam Stephanides. 1992. "Constructing Quaternions: On the Analysis of Conceptual Practice." In *Science as Practice and Culture*, edited by Andrew Pickering, 139–67. Chicago: University of Chicago Press.

Plotkin, Mark J. 1993. *Tales of a Shaman's Apprentice.* New York: Viking.

Porter, Gareth. 1990. "Post-Cold War Global Environment and Security." *Fletcher Forum of World Affairs* 14, no. 2:332–44.

Power, Michael. 1992. "After Calculation? Reflections on *Critique of Economic Reason* by Andre Gorz." *Accounting, Organization and Society* 17, no. 5:477–99.

Price, Martin. 1989. "Global Change: Defining the Ill-Defined." *Environment* 31, no. 8:18–24.

———. 1992. "The Evolution of Global Environmental Change." *Impact of Science on Society* 166:171–82.

Pugh, Emerson W. 1984. *Memories That Shaped an Industry.* Cambridge, MA: MIT Press.

Rabinow, Paul. 1992. "Artificiality and Enlightenment from Sociobiology to Biosociality." In *Incorporations*, edited by Jonathan Crary and Sanford Kwinter, 234–52. *Zone*, vol. 6. Cambridge, MA: MIT Press.

————. 1996. *Making PCR: A Story of Biotechnology*. Chicago: University of Chicago Press.

Redclift, Michael. 1992. "The Meaning of Sustainable Development." *Geoforum* 23, no. 3:395–403.

Redmond, Kent C., and Thomas M. Smith. 1980. *Project Whirlwind: The History of a Pioneer Computer*. Cambridge, MA: Digital.

Rees, Mina. 1982. "The Computing Program of the Office of Naval Research, 1946–1953." *Annals of the History of Computing* 4, no. 2:102–20.

Reintjes, J. Francis. 1991. *Numerical Control: Making a New Technology*. New York: Oxford University Press.

Repetto, Robert. 1992. "Accounting for Environmental Assets." *Scientific American* 266, no. 6 (June): 94–100.

Rheinberger, Hans-Jorg. 1995. "Beyond Nature and Culture: A Note on Medicine in the Age of Molecular Biology." *Science in Context* 8, no. 1:249–63.

————. 1997. *Towards a History of Epistemic Things*. Stanford, CA: Stanford University Press.

Richardson, George P. 1991. *Feedback Thought in Social Science and Systems Theory*. Philadelphia: University of Pennsylvania Press.

Richardson, John M. 1987. "Global Modeling: A Retrospective." *Futures Research Quarterly* 3, no. 1 (Spring): 5–26.

Rider, William H. 1992. "Operations Research and Game Theory: Early Connections." In *Toward a History of Game Theory*, edited by Roy E. Weintraub, 225–40. Durham, NC: Duke University Press.

Rolfe, J. M., and K. J. Staples. 1986. *Flight Simulation*. New York: Cambridge University Press.

Sachs, Wolfgang. 1994. "The Blue Planet: An Ambiguous Modern Icon." *Ecologist* 24, no. 5 (September–October): 170–75.

Sandbach, Francis. 1978a. "The Rise and Fall of the *Limits to Growth* Debate." *Social Studies of Science* 8:495–520.

————. 1978b. "Ecology and the Limits to Growth Debate." *Antipode* 10, no. 2: 22–32.

Sapolsky, Harvey M. 1990. *Science and the Navy*. Princeton, NJ: Princeton University Press.

Schneider, Stephen H. 1989. "The Changing Climate." *Scientific American* 261, no. 3: 70–79.

————, ed. 1992. *Gaia and the Scientists*. Cambridge, MA: MIT Press.

Serres, Michel. 1992. "The Natural Contract." *Critical Inquiry* 19:1–21.

Servan-Schreiber, Jean-Jacques. 1968. *The American Challenge*. London: Hamilton.

Smith, Bruce L. R. 1966. *The RAND Corporation*. Cambridge, MA: Harvard University Press.

Smith, James A. 1991. *The Idea Brokers*. New York: Free Press.

Solow, Herbert. 1951. "Operations Research." *Fortune* 43, no. 4 (April): 105–7, 146–48.

Steinhardt, Jacinto. 1946. "The Role of Operations Research in the Navy." *Naval Proceedings* 72, no. 3:649–56.

Taylor, James R., and Elizabeth J. Van Every. 1993. *The Vulnerable Fortress: Bureaucratic Organization and Management in the Information Age.* Toronto: University of Toronto Press.

Taylor, Peter J. 1988. "Technocratic Optimism, H. T. Odum, and the Partial Transformation of Ecological Metaphor after World War II." *Journal of the History of Biology* 21, no. 2:213–44.

Taylor, Peter J., and Frederick H. Buttel. 1992. "How Do We Know We Have Global Environmental Problems? Science and the Globalization of Environmental Discourse." *Geoforum* 23, no. 3:405–16.

Tobias, Michael. 1990. *Voice of the Planet.* New York: Bantam.

Trefethen, Florence. 1954. "A History of Operations Research." In *Operations Research for Management,* edited by Joseph McCloskey and Florence Trefethen. Vol. 1. Baltimore, MD: Johns Hopkins University Press.

Turner, Terence. 1993. "The Role of Indigenous Peoples in the Environmental Crisis: The Example of the Kayapo of the Brazilian Amazon." *Perspectives in Biology and Medicine* 36, no. 3 (Spring): 526–45.

UN Commission on Environment and Development. 1987. *Our Common Future.* New York: Oxford University Press.

Unger, Irwin, and Debi Unger. 1988. *Turning Point: 1968.* New York City: Scribner's.

U.S. Environmental Protection Agency. 1991. *Reducing Risk: Setting Priorities and Strategies for Environmental Protection.* Washington, DC: U.S. Environmental Protection Agency.

U.S. Global Change Research Program. 1990. *Our Changing Planet: The FY Research Plan.* Washington, DC: U.S. Printing.

U.S. House Committee on Science, Space, and Technology. 1990. *Participation in the International Institute for Applied Systems Analysis (IIASA).* 101st Cong., 2d sess. Washington, DC: U.S. Government Printing Office.

Vernadsky, Vladimir. 1929. *La Biosphere.* Paris: Librairie Felix Alcan.

———. 1945. "Problems of Biogeochemistry, II." *Transactions of the Connecticut Academy of Arts and Sciences* 35:483–517.

Virilio, Paul. 1989. *War and Cinema: The Logistics of Perception.* Translated by Patrick Camiller. London: Verso.

Wark, McKenzie. 1990. "The Logistics of Perception." *Meanjin* 49, no. 1 (Autumn): 95–101.

White, Frank. 1987. *The Overview Effect.* Boston: Houghton Mifflin.

Whitmore, William F. 1953. "Edison and Operations Research." *Journal of the Operations Research Society of America* 1, no. 2:83–85.

Wildes, Karl L., and Nilo A. Lindgren. 1985. *A Century of Electrical Engineering and Computer Science at MIT: 1882–1982.* Cambridge, MA: MIT Press.

Wilson, E. O. 1989. "Threats to Biodiversity." *Scientific American* 261, no. 3:108–17.

Worster, Donald. 1987. "The Vulnerable Earth: Toward a Planetary History." *Environmental Review* 11:87–103.

Archival Sources

Aurelio Peccei Papers, Ente per le Nouve Tecnologie, l'Energia e l'Ambiente, Rome.
Gordon Stanley Brown Papers, MC24, Institute Archives and Special Collections, MIT Libraries, Cambridge, MA.
Philip McCord Morse Papers, MC75, Institute Archives and Special Collections, MIT Libraries, Cambridge, MA.
Carroll Louis Wilson Papers, MC29, Institute Archives and Special Collections, MIT Libraries, Cambridge, MA.

Interviews

Umberto Colombo, interview by the author, Rome, Italy, October 19, 1992.
Ricardo Diez-Hochleitner, interview by the author, by phone, April 1, 1993.
Jay Forrester, interviews by the author, Cambridge, MA, December 12, 1991, and May 20, 1994.
Eleonora Masini, interview by the author, Rome, Italy, November 23, 1992.
Dennis Meadows, interview by the author, by phone, May 31, 1994.
Hasan Ozbekhan, interview by the author, Philadelphia, PA, May 27, 1994.
Anna Pignocchi, interview by the author, Rome, Italy, November 11, 1992.

Index

Abelson, Philip, 106
Accademia Nazionali del Lincei, 64, 70
Acid rain, 16
ADELA (Atlantic Development of Latin America), 62
Advertising linked to planetary concerns, 10–11
Africa: Fellows in (MIT program), 82
Agnelli family, 64, 74n15
Agriculture, 96; food production needs and policies, 108, 109; impact of climatic changes studied, 93n23
Air Defense System (ADS), 35
Air Force, U.S., 49; Defense Meteorological Satellite Program, 9; High Command, 61
Airplane Simulator and Control Analyzer (ASCA), 42–43, 47, 48, 55
Alternatives to Growth (Meadows, ed.), 105
American Academy of Arts and Sciences, 92n21
American Challenge, The (Servan-Schreiber), 68
Amoco Corporation advertising, 11
Anthropocentric position, shift from, 21–22
Antisubmarine warfare (World War II), 24–25, 27, 28, 29–30, 31
Antisubmarine Warfare Operations Research Group, 29
Apollo 8 space mission, 7
Apollo 9 space mission, 8
Atlantic Development of Latin America (ADELA), 62
Atomic Energy Commission, 36, 82
Austria, and Club of Rome, 92n11

Baker, James, 13–14

Bamberger stores, 29
Barney, Gerald, 88
Battelle Institute (Geneva), 65, 79–80
Battle of Britain. *See* World War II
Baum, Claude, 65
Baxter, James P., 26, 29–30
Beeline Expressway (Florida), and destruction of habitat, 23n2
Bell Telephone Laboratories, 29
Bergman, Charles, 8
Beyond Interdependence: The Meshing of the World's Economy and the Earth's Ecology (MacNeill et al.), 15
Beyond the Limit (Meadows, Meadows, and Randers), 110–11
Biodiversity, loss of, 10
Biofera (Vernadsky), 17
Biosphere: concept emerges, 17–19, 22; and Gaia hypothesis, 17, 18–21; and human survival, 21; Vernadsky's conception of, 17–18, 20, 21
Birth control, 96, 109. *See also* Population growth
Blackett, P. M. S., and "Blackett's Circus," 26, 27–28
Boeing Corporation advertising, 10–11
Bowker, Geoff, 45
Bowles, Edward L., 50
Britain. *See* Great Britain
Bronk, Detlev, 75
Brown, Gordon Stanley, 41, 42, 46, 50, 73n6, 88
Brown, Lester, 106, 114n23
Bulletin of the European Economic Community, 63
Bundy, McGeorge, 84–85